Eliot Attridge

REVISION PLUS

OCR Twenty First Century
GCSE Biology

Rev

Companion

Ideas about Science

Introduction to Ideas about Science

The OCR Twenty First Century Biology specification aims to ensure that you develop an understanding of science itself – of how scientific knowledge is obtained, the kinds of evidence and reasoning behind it, its strengths and limitations, and how far we can rely on it.

These issues are explored through Ideas about Science, which are built into the specification content and summarised over the following pages.

The tables below give an overview of the Ideas about Science that can be assessed in each unit and provide examples of content which support them in this guide.

Unit A161 (Modules B1, B2 and B3)

Ideas about Science	Example of Supporting Content
Cause–effect explanations	Variation (page 3)
Developing scientific explanations	Evolution (page 28)
The scientific community	Evolution (page 28)
Risk	Risks (page 13)
Making decisions about science and technology	Reliability (page 9)

Unit A162 (Modules B4, B5 and B6)

Ideas about Science	Example of Supporting Content
Data: their importance and limitations	Fieldwork (page 37)
Cause–effect explanations	Limitations to Photosynthesis (page 36)
Developing scientific explanations	Memory Models (page 57)
Making decisions about science and technology	Ethical Decisions (page 48)

Unit A163 (Module B7)

Ideas about Science	Example of Supporting Content
Data: their importance and limitations	Body Mass Index (page 63)
Cause–effect explanations	Diabetes (page 71)
Developing scientific explanations	Human Activity – Removing Resources (page 76)
The scientific community	Closed Loop System Diagrams (page 73)
Risk	Stem Cell Technology (page 83)
Making decisions about science and technology	Could We, Should We? (page 78)

❶ Data: Their Importance and Limitations

Science is built on data. Biologists carry out experiments to collect and interpret data, seeing whether the data agree with their explanations. If the data do agree, then it means the current explanation is more likely to be correct. If not, then the explanation has to be changed.

Experiments aim to find out what the 'true' value of a quantity is. Quantities are affected by errors made when carrying out the experiment and random variation. This means that the measured value may be different to the true value. Biologists try to control all the factors that could cause this uncertainty.

Biologists always take repeat readings to try to make sure that they have accurately estimated the true value of a quantity. The mean is calculated and is the best estimate of what the true value of a quantity is. The more times an experiment is repeated, the greater the chance that a result near to the true value will fall within the mean.

The range, or spread, of data gives an indication of where the true value must lie. Sometimes a measurement will not be in the zone where the majority of readings fall. It may look like the result (called an 'outlier') is wrong – however, it does not automatically mean that it is. The outlier has to be checked by repeating the measurement of that quantity. If the result cannot be checked, then it should still be used.

Here is an example of an outlier in a set of data:

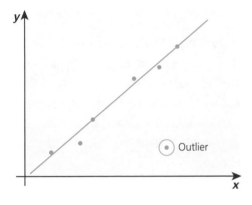

Outlier

The spread of the data around the mean (the range) gives an idea of whether it really is different to the mean from another measurement. If the ranges for each mean do not overlap, then it is more likely that the two means are different. However, sometimes the ranges do overlap and there may be no significant difference between them.

The ranges also give an indication of reliability – a wide range makes it more difficult to say with certainty that the true value of a quantity has been measured. A small range suggests that the mean is closer to the true value.

If an outlier is discovered, you need to be able to defend your decision as to whether you keep it or discard it.

❷ Cause–effect Explanations

Science is based on the idea that a factor has an effect on an outcome. Biologists make predictions as to how the input variable will change the outcome variable. To make sure that only the input variable can affect the outcome, biologists try to control all the other variables that could potentially alter it. This is called 'fair-testing'.

You need to be able to explain why it is necessary to control all the factors that might affect the outcome. This means suggesting how they could influence the outcome of the experiment.

A correlation is where there is an apparent link between a factor and an outcome. It may be that as the factor increases, the outcome increases as well. On the other hand, it may be that when the factor increases, the outcome decreases.

For example, in plants, there is a correlation that up to 40°C, the higher the temperature, the greater the rate of photosynthesis.

Just because there is a correlation does not necessarily mean that the factor causes the outcome. Further experiments are needed to establish this. It could be that another factor causes the outcome or that both the original factor and outcome are caused by something else.

The following graph suggests a correlation between going to the opera regularly and living longer. It is far more likely that if you have the money to go to the opera, you can afford a better diet and health care. Going to the opera is not the true cause of the correlation.

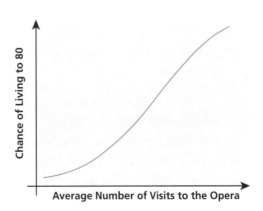

Sometimes the factor may alter the chance of an outcome occurring but does not guarantee it will lead to it. The statement 'the more time spent on a sun bed the greater the chance of developing skin cancer' is an example of this type of correlation, as some people will not develop skin cancer even if they do spend a lot of time on a sun bed.

To investigate claims that a factor increases the chance of an outcome, biologists have to study groups of people who either share as many factors as possible or are chosen randomly to try to ensure that all factors will present in people in the test group. The larger the experimental group, the more confident biologists can be about the conclusions made.

 Even so, a correlation and cause will still not be accepted by biologists unless there is a scientific mechanism that can explain them.

❸ Developing Scientific Explanations

Biologists devise hypotheses (predictions of what will happen in an experiment), along with an explanation (the scientific mechanism behind the hypotheses) and theories (that can be tested).

Explanations involve thinking creatively to work out why data have a particular pattern. Good scientific explanations account for most or all of the data already known. Sometimes they may explain a range of phenomena that were not previously thought to be linked. Explanations should enable predictions to be made about new situations or examples.

When deciding on which is the better of two explanations, you should be able to give reasons why.

Explanations are tested by comparing predictions based on them with data from observations or experiments. If there is an agreement between the experimental findings, then it increases the chance of the explanation being right. However, it does not prove it is correct. Likewise, if the prediction and observation indicate that one or the other is wrong, it decreases the confidence in the explanation on which the prediction is based.

❹ The Scientific Community

Once a biologist has carried out enough experiments to back up his/her claims, they have to be reported. This enables the scientific community to carefully check the claims, something which is required before they are accepted as scientific knowledge.

Biologists attend conferences where they share their findings and sound out new ideas and explanations. This can lead to biologists revisiting their work or developing links with other laboratories to improve it.

The next step is writing a formal scientific paper and submitting it to a journal in the relevant field. The paper is allocated to peer reviewers (experts in their field), who carefully check and evaluate the paper. If the peer reviewers accept the paper, then it is published. Biologists then read the paper and check the work themselves.

New scientific claims that have not been evaluated by the whole scientific community have less credibility than well-established claims. It takes time for other biologists to gather enough evidence that a theory is sound. If the results cannot be repeated or replicated by themselves or others, then biologists will be sceptical about the new claims.

If the explanations cannot be arrived at from the available data, then it is fair and reasonable for different biologists to come up with alternative explanations. These will be based on the background and experience of the biologists. It is through further experimentation that the best explanation will be chosen.

This means that the current explanation has the greatest support. New data are not enough to topple it. Only when the new data are sufficiently repeated and checked will the original explanation be changed.

 You need to be able to suggest reasons why an accepted explanation will not be given up immediately when new data, which appear to conflict with it, have been published.

⑤ Risk

Everything we do (or not do) carries risk. Nothing is completely risk-free. New technologies and processes based on scientific advances often introduce new risks.

Risk is sometimes calculated by measuring the chance of something occurring in a large sample over a given period of time (calculated risk). This enables people to take informed decisions about whether the risk is worth taking. In order to decide, you have to balance the benefit (to individuals or groups) with the consequences of what could happen.

For example, deciding whether or not to have a vaccination involves weighing up the benefit (of being protected against a disease) against the risk (of side effects).

Risk which is associated with something that someone has chosen to do is easier to accept than risk which has been imposed on them.

> **HT** Perception of risk changes depending on our personal experience (perceived risk). Familiar risks (e.g. smoking) tend to be under-estimated, whilst unfamiliar risks (e.g. a new vaccination) and invisible or long-term risks (e.g. radiation) tend to be over-estimated.
>
> For example, many people under-estimate the risk of getting type 2 diabetes from eating too much unhealthy food.

Governments and public bodies try to assess risk and create policy on what is and what is not acceptable. This can be controversial, especially when the people who benefit most are not the ones at risk.

⑥ Making Decisions about Science and Technology

Science has helped to create new technologies that have improved the world, benefiting millions of people. However, there can be unintended consequences of new technologies, even many decades after they were first introduced. These could be related to the impact on the environment or to the quality of life.

When introducing new technologies, the potential benefits must be weighed up against the risks.

Sometimes unintended consequences affecting the environment can be identified. By applying the scientific method (making hypotheses, explanations and carrying out experiments), biologists can devise new ways of putting right the impact. Devising life cycle assessments helps biologists to try to minimise unintended consequences and ensure sustainability.

Some areas of biology could have a high potential risk to individuals or groups if they go wrong or if they are abused. In these areas the Government ensures that regulations are in place.

The scientific approach covers anything where data can be collected and used to test a hypothesis. It cannot be used when evidence cannot be collected (e.g. it cannot test beliefs or values).

Just because something can be done does not mean that it should be done. Some areas of scientific research or the technologies resulting from them have ethical issues associated with them. This means that not all people will necessarily agree with it.

Ethical decisions have to be made, taking into account the views of everyone involved, whilst balancing the benefits and risks.

It is impossible to please everybody, so decisions are often made on the basis of which outcome will benefit most people. Within a culture there will also be some actions that are always right or wrong, no matter what the circumstances are.

Contents

Module B1 (You and Your Genes)

Many of an individual's characteristics are inherited from their two biological parents. This module looks at:
- genes and their effect on development
- how understanding genetic information can be used to prevent disease
- how genetic information can and should be used.

Genetic Information

All organisms develop following a set of instructions that are coded inside the cell in the nucleus. The instructions control how the organism develops and functions. The basic unit for the instructions is called the **gene**. Genes occur in very long **DNA** (deoxyribonucleic acid) molecules called **chromosomes**.

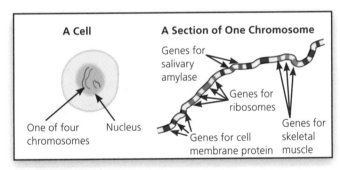

A Cell

A Section of One Chromosome

Genes for salivary amylase

Genes for ribosomes

Genes for skeletal muscle

Genes for cell membrane protein

One of four chromosomes

Nucleus

Chromosomes are made of DNA molecules. Each DNA molecule consists of two strands, which form a **double helix**.

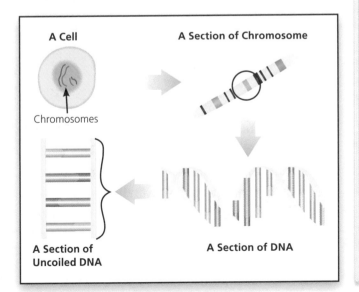

A Cell

A Section of Chromosome

Chromosomes

A Section of DNA

A Section of Uncoiled DNA

Genes are sections of DNA that describe how to make proteins. These may be structural (e.g. for collagen in skin) or functional (e.g. for enzymes such as amylase, which breaks down starch). Some characteristics are coded for by a number of genes that work together, e.g. eye colour.

Eye Colour

Originally it was believed that eye colour was due to a single gene. It is now known that there are a number of genes coding for the different pigments in the iris, mainly on chromosome 15 in humans. This means that there is an enormous variation in eye colour.

Variation

The differences between individuals of the same species are described as **variations**.

Variation may be due to:
- **genes** – the different characteristics that an individual inherits, e.g. whether you have dimples or not
- **environmental factors** – how the environment changes an individual, e.g. cutting the skin may cause a scar.

Dimple

Scar

Variation is usually due to a combination of genes and the environment, e.g. your weight. Biologists carry out a lot of research to try to determine whether a characteristic is genetic or environmental. Even if a gene is discovered, it does not mean that it is the only factor in that characteristic.

> **HT** **Genotype** is the term describing the genetic makeup of an organism (the combination of alleles). **Phenotype** describes the observable characteristics the organism has.

Alleles

All body cells contain pairs of chromosomes. The genes in each chromosome are in the same place on each one, which means that body cells usually have two copies of each gene. These copies of each gene can be different versions, called **alleles**.

Alleles are described as being either **dominant** or **recessive**. A dominant allele is one that controls the development of a characteristic even if it is present on only one chromosome in a pair. A recessive allele controls the development of a characteristic only if the dominant allele is not present, i.e. both chromosomes have the recessive allele present.

The allele of a gene is usually represented by a letter. If the allele is dominant, it is denoted by a capital letter. If the allele is recessive, it is denoted by a lower-case letter. As each body cell has two alleles for each gene, they can be the same or different.

For example, the ability to roll your tongue is dominant so it can be represented by T. If your alleles were TT or Tt, then you would be able to roll your tongue. If your alleles were tt, you would not be able to roll your tongue.

If you carried one of each allele then you would not express the recessive characteristic. You would be a carrier for that allele but would have the dominant allele expressed.

> When the two alleles are the same we say that they are **homozygous**. When they are different they are **heterozygous**. For example, TT = homozygous dominant; tt = homozygous recessive; Tt = heterozygous.

Sex Cells

The **sex cells**, eggs produced by the **ovaries** in females and sperm produced by the **testes** in males, only carry one copy of each chromosome. This is the basis of sexual reproduction.

Humans have 23 pairs of chromosomes in their body cells (46 in total). The sex cells have half the amount, i.e. 23 single chromosomes.

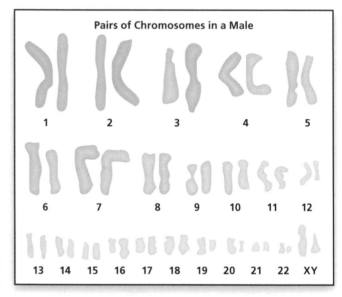

Pairs of Chromosomes in a Male

1 2 3 4 5
6 7 8 9 10 11 12
13 14 15 16 17 18 19 20 21 22 XY

This means that when fertilisation takes place (i.e. the nuclei of the sperm and the egg fuse), the total number of chromosomes doubles as they pair up again.

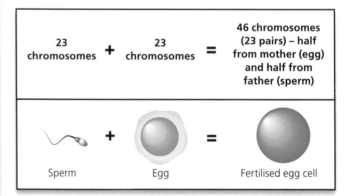

23 chromosomes **+** 23 chromosomes **=** 46 chromosomes (23 pairs) – half from mother (egg) and half from father (sperm)

Sperm **+** Egg **=** Fertilised egg cell

As the pairing up of the chromosomes is random, the new offspring will differ from its parents. This leads to variation, a major advantage of sexual reproduction. The child will share similarities with its parents depending on which characteristics have come from the father, which have come from the mother and which ones are dominant and recessive. The child will also differ from any brothers and sisters.

Genetic Diagrams

It is easiest to follow what is happening with the inheritance of gene characteristics by drawing genetic diagrams.

Family trees can be used to trace the inheritance of a characteristic and to work out who must have been carrying a faulty allele.

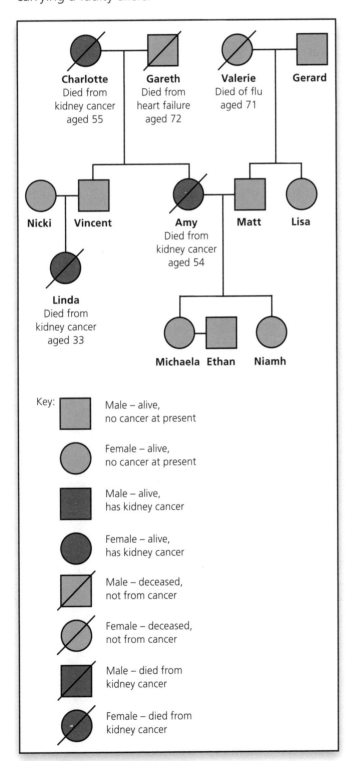

Key:

▢	Male – alive, no cancer at present
○	Female – alive, no cancer at present
■	Male – alive, has kidney cancer
●	Female – alive, has kidney cancer
▨	Male – deceased, not from cancer
⊘	Female – deceased, not from cancer
▨	Male – died from kidney cancer
⊘	Female – died from kidney cancer

As Amy had kidney cancer, it is possible she carried both recessive alleles. Michaela would be worried as she would suspect that she may have inherited one of the alleles from Amy. As the cancer must be due to a homozygous recessive allele, she would be correct. If Matt was also a carrier, then Michaela would have the chance that she has both alleles causing the cancer.

When looking at the possibilities of inheriting an allele, we can use a **Punnett square diagram**. This shows all the possible pairings of alleles from sperm and egg at fertilisation.

With a Punnett square, the possible versions of sperm and egg are placed at the sides and the possible offspring that could result are plotted.

For example, if a male with a dominant **A** allele and recessive **a** allele was to mate with a female with the same alleles, the following diagram could be drawn:

		♂	
		A	**a**
♀	**A**	AA	Aa
	a	Aa	aa

This means that three of the four possible offspring would show the dominant characteristic while only one of the four possible offspring would be recessive for both alleles.

HT The diagram shows that the possible genotypes would be one homozygous dominant offspring, two heterozygous offspring and one homozygous recessive offspring.

The genotypes would be in the ratio 1 : 2 : 1. The ratio of the phenotypes would be 3 : 1 (three dominant to one recessive).

Sex Chromosomes

One of the 23 pairs of chromosomes in a human cell is the sex chromosome. In females the sex chromosomes are the same – they are both X chromosomes. In males they are different – there is an X chromosome and a Y chromosome. The Y chromosome is much shorter than the X chromosome.

We can draw a Punnett square to represent how sex is determined. This time, rather than an allele, we are writing the whole chromosome that will be carried by the sperm or egg (X or Y).

		♂	
		X	**Y**
♀	**X**	XX	XY
	X	XX	XY

Therefore, 50% of the offspring will be female and 50% male. As the process of fertilisation is completely random, some families will only have girls whilst others will only have boys.

Rare Disorders

Most characteristics are governed by a range of genes, so the presence of one 'faulty' allele may not affect the overall outcome.

Usually, disorders are caused by a recessive gene, but occasionally the faulty gene is dominant, meaning that only one allele needs to be present for the disorder to be expressed.

For example, **Huntington's disease** is caused by the presence of a single dominant faulty allele. This means that if a parent carries the dominant allele, then the child has a 50% chance of carrying it too.

Cystic fibrosis, on the other hand, is caused by the presence of a faulty recessive allele. Both recessive alleles are needed for the disease to develop.

Sex Determination

The sex of an embryo is determined by a gene on the Y chromosome called the SRY (sex-determining region Y) gene.

If the gene is not present, i.e. if there are two X chromosomes present, the embryo will develop into a female and ovaries will grow. If the gene is present, i.e. if both an X and a Y chromosome are present, then testes will begin to develop.

Six weeks after fertilisation, the undifferentiated gonads start producing a hormone called **androgen**. Specialised receptors in the developing embryo detect the androgen. This stimulates the male reproductive organs grow.

Sometimes the Y chromosome is present, but the androgen is *not* detected. When this happens, the embryo develops all the female sex organs except the uterus. The baby is born with a female body but will be infertile.

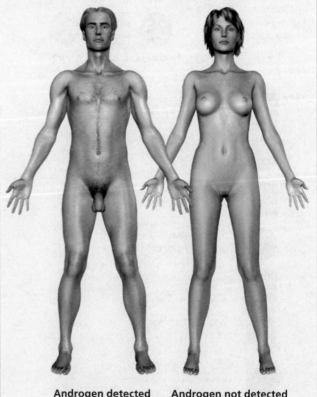

Androgen detected
Genetically male
Appears male

Androgen not detected
Genetically male
Appears female; no uterus

Huntington's Disease

Huntington's disease is a genetic disorder that affects the central nervous system. It is caused by a faulty dominant allele on chromosome 4.

The allele that causes the disease results in damage to nerve cells in certain areas of the brain. This leads to gradual physical, mental and emotional changes that are expressed as symptoms.

The symptoms of the disease normally develop in adulthood, which means sufferers may have already had children and passed on the gene. Symptoms include late onset, a tremor, clumsiness, memory loss, an inability to concentrate and mood changes.

Everyone who inherits the Huntington's allele will develop the disease. This is due to the allele being dominant. Only one parent needs to pass on the allele for a child to develop the disorder.

♂

	H	h
h	Hh	hh
h	Hh	hh

♀

50% have Huntington's disease, 50% do not.

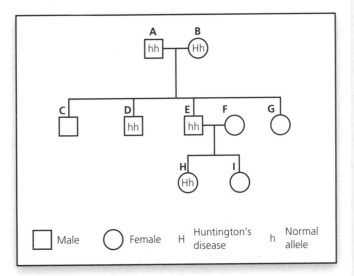

| Male | Female | H | Huntington's disease | h | Normal allele |

With individual I there is a 50% chance of inheriting the H allele if the mother (F) was Hh and 100% if she was HH.

Cystic Fibrosis

Cystic fibrosis is the UK's most common life-threatening genetic disease. It affects the cell membranes, causing a **thick mucus** to be produced in the lungs, gut and pancreas.

Other symptoms of cystic fibrosis are difficulty in breathing, an increased number of chest infections and difficulty in digesting food.

Although there is no cure at present, scientists have identified the faulty recessive allele that causes cystic fibrosis and are looking at ways to repair or replace it.

Unlike Huntington's disease, the cystic fibrosis allele is recessive. Therefore, if an individual is a carrier (has only one recessive allele) they will not have the characteristics of the disease. They can, however, pass on the allele to their children.

♂

	F	f
F	FF	Ff
F	FF	Ff

♀

50% are carriers of the cystic fibrosis allele.

♂

	F	f
F	FF healthy	Ff carrier
f	Ff carrier	ff CF

♀

25% are healthy, 50% are carriers and 25% will have cystic fibrosis.

Genetic Testing

It is now possible to test adults, children and embryos for a faulty allele if there is a family history of a genetic disorder. If the test turns out positive, the individual will have to decide whether or not to have children and risk passing on the disorder. This is called **predictive testing for genetic diseases**.

Genetic testing can also be carried out to determine whether an adult or child can be prescribed a particular drug without suffering from serious side effects.

For example, certain people are highly susceptible to getting liver damage while taking COX-2 (an enzyme in the body) inhibitor drugs. A genetic test would ensure that *only* those patients who do *not* have the susceptibility gene are prescribed the drug.

Embryos can be tested for embryo selection. The healthy embryos that do not have the faulty allele are then implanted. This process involves harvesting egg cells from the mother and then fertilising them with the father's sperm. Only the healthy embryos are implanted into the mother's uterus, where the pregnancy progresses as normal. This process is called *in vitro* **fertilisation** (IVF).

HT The procedure for embryo selection is called **pre-implantation genetic diagnosis** (PGD). After fertilisation, the embryos are allowed to divide into eight cells before a single cell is removed from each one for testing. The selected cell is then tested to see if it carries the alleles for a specific genetic disease, i.e. the allele for the disease that one of the parents carries.

The PGD test is not without risk – the result may be inaccurate and lead to a healthy embryo not being implanted. It may also decrease the chance of the embryo surviving once it has been implanted.

Risks of Genetic Testing

Testing is not without risk. When a fetus is tested, there is a risk that the test itself could cause the death of the fetus. There are two ways to carry out a test on a developing fetus:

1 Amniocentesis Testing

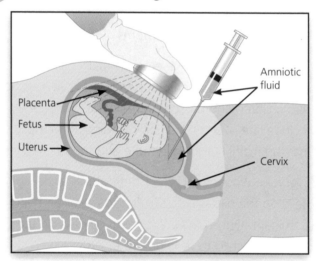

The amniocentesis test can be carried out at 14–16 weeks of pregnancy. A needle is inserted into the uterus, taking care to avoid the fetus itself, and a small sample of amniotic fluid, which carries cells from the fetus, is extracted.

If the test is positive for a given disease, then the pregnancy (now at 16–18 weeks) could be terminated. There is approximately a 0.5% (1 in 200) chance of the test itself causing a miscarriage. There is also a very small chance of infection.

2 Chorionic Villus Testing

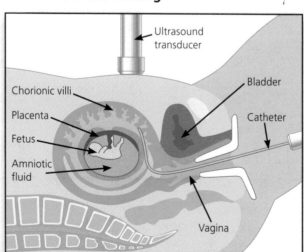

Chorionic villus sampling can take place earlier in pregnancy, at 8–10 weeks. A special catheter is inserted through the vagina and cervix until it reaches the placenta. Part of the placenta has finger-like protrusions called chorionic villi. Samples are removed for testing.

If the test is positive for the faulty allele then the pregnancy can be terminated much earlier (10–12 weeks). However, the chance of a miscarriage is much higher at approximately 2% (1 in 50).

Reliability

Because no test is 100% reliable, genetic testing can have a number of possible outcomes:

Outcome	Test Result	Reality
True positive	Subject has the disorder	Subject has the disorder
True negative	Subject does not have the disorder	Subject does not have the disorder
False positive	Subject has the disorder	Subject **does not** have the disorder
False negative	Subject does not have the disorder	Subject **has** the disorder

If the subject was a fetus, then the consequences of a false positive result could mean that the pregnancy was terminated, leading to the death of the fetus. On the other hand, a false negative result could lead to the birth of a baby with a severe genetic disease.

If the subject was a parent deciding whether or not to have children, the consequence of a false positive result could mean that the parent decides not to have children.

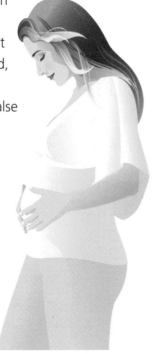

With all genetic tests, decisions also have to be taken as to whether other members of the family should be told of the test and the result. Although an individual might be willing to take a genetic test, other family members may not want to know.

For example, if a parent had a positive test for Huntington's disease, then there is at least a 50% chance that a child will also carry the allele. Should the child be told? Huntington's disease occurs late in life and is incurable. Is it fair to make someone worry about the condition decades before they are likely to get it?

These are **ethical considerations** and need to be considered carefully before decisions are taken.

There is always a difference between what *can* be done (i.e. what is technically possible) and what *should* be done (i.e. what is morally acceptable).

For example, governments may have the ability to impose genetic tests on individuals by implementing genetic screening programmes, but should they be allowed to do so?

There is the potential for genetic testing to be used to produce detailed genetic profiles. These could contain information on everything from an individual's ethnicity to whether they are susceptible to certain conditions (e.g. obesity) or diseases (e.g. cancer).

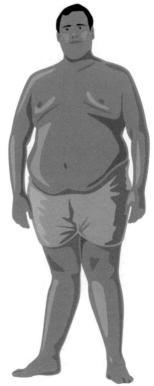

The question is: how will the information be used? Employers could potentially refuse to employ someone who possessed certain alleles and insurers may not cover a person who had genes that made them more likely to suffer a heart attack. Would this be fair?

B1 | You and Your Genes

Asexual Reproduction

When a cell grows and divides into two, it is a form of reproduction. As it does not involve sex, it is called **asexual reproduction**.

All bacteria reproduce asexually, as do many plants and some animals. As they are formed by the 'mother' cell dividing to form two 'daughter' cells, the genes in each cell are exactly the same. When an organism has exactly the same genetic information as another individual, it is called a **clone**.

The only differences between clones are due to the environment. For example, if a cloned plant received more water and sunlight than another, it would grow better.

Plants such as strawberries produce shoots called **runners**. These eventually break off and become new strawberry plants, clones of the original.

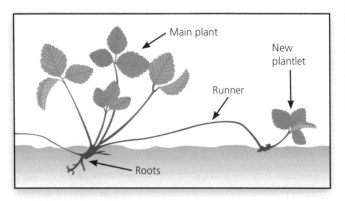

Main plant

New plantlet

Runner

Roots

Other plants grow **bulbs**. When bulbs are planted they grow into genetically identical plants. Again the environment will alter them. No two organisms can occupy the same space in the universe so the environment will always be different for individuals, even if they are clones.

Clones of animals occur naturally when, during the earliest stages after fertilisation, the developing embryo splits into two. This leads to the creation of **identical twins**.

Identical triplets are also possible, although they are extremely rare. They occur when the fertilised egg splits into two and one of the new cells splits into two again.

It is now possible to make clones artificially by taking the nucleus from an adult body cell and transferring it into an empty, unfertilised egg cell. The process has been successful in a wide range of organisms, the most famous of which was a sheep named Dolly.

Stem Cells

Cloning depends on cells that have the potential to become any cell type in the body. These are called **stem cells**.

Adult stem cells are unspecialised cells that can develop into many, but not all, types of cell.

Embryonic stem cells are unspecialised cells that can develop into *any* type of cell, including more embryonic stem cells.

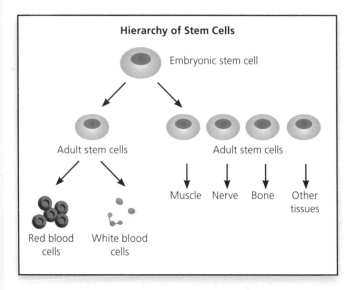

Hierarchy of Stem Cells

Embryonic stem cell

Adult stem cells

Adult stem cells

Red blood cells

White blood cells

Muscle Nerve Bone Other tissues

Both types of stem cell can be used to treat some illnesses or injuries. For example, skin can be grown as a treatment for serious burns and sight can now be restored to people who are blind due to damage of their corneas.

After the zygote has divided four times to reach the 16 cell stage, the majority of cells in the embryo start to become specialised. This means that certain genes are switched on and off. This leads to the production of proteins that are specific to the specialised cell type. Specialised cells can only divide to produce the same type of specialised cell.

Module B2 (Keeping Healthy)

To stay healthy it is important to maintain a healthy lifestyle and use medication when appropriate. This module looks at:

- how our bodies resist infection
- what vaccines are and how they work
- what antimicrobials are and why they can become less effective
- how new drugs are developed and tested
- what factors increase the risk of heart disease.

Microorganisms

Microorganisms are organisms that are too small to see with the naked eye. They include **bacteria**, **viruses** and **fungi**.

They can be beneficial to us (e.g. the bacteria that live in our intestines can produce certain vitamins) or they can cause us harm (e.g. bacteria that cause food poisoning).

Symptoms of Infectious Disease

Once inside the body, harmful microorganisms start to reproduce. As they grow in number they start to damage cells, often by bursting them (**lysis**). Sometimes they also produce toxins (poisons). When the damage to cells or the amount of toxin reaches a certain level then the symptoms of the disease will appear.

The human body provides ideal conditions for microorganisms to grow. In the body, there is water, oxygen (although not all microorganisms require this), food and heat, as well as different pH levels.

When the conditions are suitable, microorganisms can use these resources to reproduce very quickly. Some bacteria can take as little as 15 minutes to divide, meaning that they can increase rapidly in number. However, other bacteria can take days to divide.

This form of growth is known as **exponential growth**. It follows the formula:

$$x(t) = a \times b^{t/\tau}$$

where
x = the quantity of bacteria at a given time
a = amount of bacteria at start
b = growth factor
t = time
τ = time taken to double

For example, if a bacterium doubled every 20 minutes, how many would there be after an hour?

$a = 1$, $b = 2$, $t = 60$min and $\tau = 20$min

$$x(t) = a \times b^{t/\tau}$$
$$x(60\text{min}) = 1 \times 2^{60\text{min}/20\text{min}}$$
$$= 8$$

In four hours there would be 4096 bacteria; in 12 hours there would be 68 719 476 736 bacteria; and in 24 hours there would be 4.7×10^{21} bacteria.

In practice, these numbers would never be reached. The resources inside the body become less readily available as the number of bacteria increases.

A growth curve shows the realistic stages of bacterial growth:

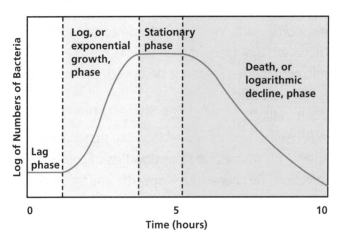

Using a soap that claimed to kill 99.9% of bacteria would reduce the original number by 1000 fold. So, if you had 8192 bacteria on your hands, washing would leave you with eight bacteria. However, they could still reproduce within a few hours to a great number.

The Immune System

The body has immune systems that defend it against invading microorganisms. These immune systems are layered, which means that as each defence is compromised there is a new layer for the microorganisms to try to break through.

Physical Barriers

In order to enter the body, microorganisms first have to breach the physical barriers. The skin is the first physical barrier – it has to be cut to allow entry. Sweat is another barrier – it has antimicrobial properties.

Other areas where microorganisms can enter include the eye (protected by chemicals in tears) and the stomach (where one of the functions of stomach acid is to sterilise the food, killing microorganisms).

The Immune Response

If microorganisms do breach the physical barriers then the immune system (the body's internal defence system) is activated. White blood cells play a major role in this response. There are a large number of white blood cell types, two of which are:

1 Neutrophils

A neutrophil is a type of white blood cell that moves around the body in the bloodstream looking for microorganisms. When it finds some, it engulfs (flows around) them. It then digests the microorganism so that it is destroyed. This behaviour is **non-specific**.

2 B-cells

A B-cell is a type of white blood cell that makes special substances called **antibodies** to combat infection. This behaviour is **specific** and leads to B-cells targeting the same organism if it invades again. They are, in effect, **memory cells**. B-cells travel mainly in the **lymphatic system**.

Antibodies and Antigens

Each microorganism has its own unique markers made out of protein on its surface. These markers are called **antigens**. The antibodies produced by

B-cells are specific to a particular antigen. For example, only TB antibodies will work with TB antigens; they will not work with the antigens from cholera.

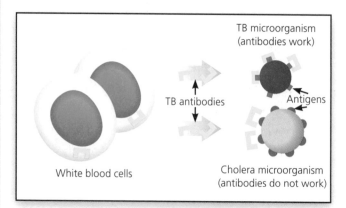

TB microorganism (antibodies work)

TB antibodies

Antigens

White blood cells

Cholera microorganism (antibodies do not work)

Once the invading microorganisms have been identified by antibodies, other white blood cells can consume them, ridding the body of the disease.

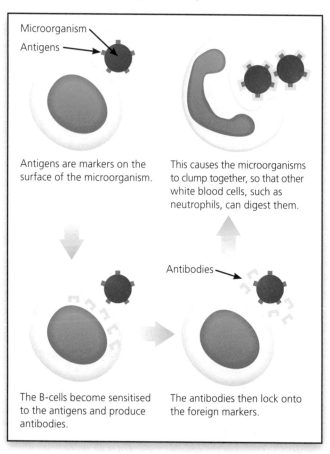

Microorganism

Antigens

Antigens are markers on the surface of the microorganism.

This causes the microorganisms to clump together, so that other white blood cells, such as neutrophils, can digest them.

Antibodies

The B-cells become sensitised to the antigens and produce antibodies.

The antibodies then lock onto the foreign markers.

The process of producing antibodies enables the immune response to be very rapid if the same microorganism infects the body again. This protects the body and gives it **immunity** against that microorganism in future.

Vaccination

Vaccination helps the body to develop long-term immunity against a disease, i.e. by producing specific antibodies. If the body is re-infected by the same microorganism, memory cells produce antibodies quickly so that the microorganism is destroyed before damage is done. This is how vaccination works:

1 Injection of vaccine

A safe form of the disease-causing microorganism is injected into the body.

2 Immune response triggered

Although the microorganism is safe, the antigens on its surface still cause the white blood cells to produce specific antibodies.

3 Memory cells remain in body

Long after the vaccination, memory cells patrol the body. If the disease-causing microorganism infects the body again, the white blood cells can attack it very quickly.

> **HT** In order to prevent an **epidemic** of a disease (like measles) in a population, it is important that as many individuals as possible are vaccinated.
>
> If more than 95% of the population is vaccinated then the unvaccinated will be protected too. This is because the risk of coming into contact with an infected person will be very small. If the percentage drops below 95%, unvaccinated individuals are more likely to get the disease and pass it on to others.
>
> If many people have the disease, the microorganism has a greater chance of mutating owing to the large number of carriers. In this case, even the vaccinated people will no longer be immune as the vaccine will be for the old form.

Risks

There is no guarantee that all vaccines and drugs (medicines) are risk free. People have genetic differences, so they may react to a vaccine or a drug in different ways. These are called **side effects**.

People always have to balance the side effects with the risk of getting the actual disease. Most side effects are minor, e.g. a mild fever or a rash.

More **extreme** side effects, e.g. encephalitis (inflammation of the brain) or convulsions, are **rare**. With the MMR (Mumps, Measles and Rubella) vaccination, the chance of getting encephalitis is 1 in 1 000 000. The risk of getting it from measles itself is between 1 in 200 and 1 in 5000 – much higher.

Some people have genes that predispose them to getting a particular side effect. For example, COX-2 Inhibitors (a type of drug used for pain relief) can cause liver damage in susceptible individuals.

Antimicrobials

Antimicrobials are chemicals that kill bacteria, fungi and viruses. An example of an antimicrobial is the metal silver, which kills bacteria.

> **HT** Antimicrobials are also used to describe chemicals that inhibit the growth of microorganisms.

Antibiotics are chemicals that are only effective against bacteria. Antibiotics are not effective against viruses, which is why you are not given them when you have the common cold or influenza.

Antimicrobial Resistance

Over a period of time, bacteria can become resistant to antimicrobials.

> **HT** Mutations (random changes) can take place in the genes of microorganisms. This leads to new strains of bacteria and fungi that are no longer affected by the antimicrobial. These reproduce and pass on the resistance. As a result, the antimicrobial is no longer effective.

To prevent resistance to antibiotics increasing:
- doctors should only prescribe them when completely necessary
- patients should always complete a course of antibiotics, even if they are feeling better.

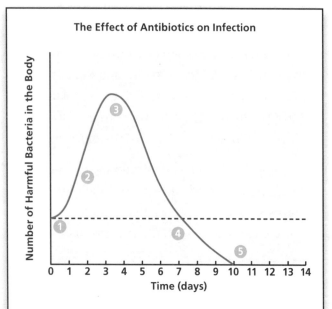

The Effect of Antibiotics on Infection

(Graph: Number of Harmful Bacteria in the Body vs Time (days), 0 to 14)

1. Harmful bacteria enter the body (by food poisoning).
2. Bacteria multiply. Patient begins to feel unwell.
3. Patient visits doctor. Starts taking antibiotics.
4. Number of bacteria now lower than originally entered the body. Patient feels better (but bacteria not all dead).
5. All harmful bacteria now destroyed.

Testing Drugs and Vaccines

Scientists are always trying to develop new drugs to fight infection. Before they can be used it is essential that the drugs are tested for **safety** and for **effectiveness**. The methods used can be controversial.

Tests on Different Types of Human Cells Grown in the Laboratory
Advantages
• Shows if drugs and vaccines are effective at targeting the problem at the cell level.
• Shows if drugs will cause damage to cells.
• No people or animals are harmed.
Disadvantages
• Does not show effects of drugs on the whole organism.
• Some people believe that growing human cells in this way is unnatural or wrong.

Tests on Animals
Advantages
• Shows if drugs are effective within body conditions.
• Shows if drugs are safe for the whole body.
Disadvantages
• Animals can suffer and die as a result of the tests.
• Animals might react differently to humans.

Following these initial tests (which can take years) **clinical trials** are carried out on **healthy volunteers** to test for safety, and on people **with the illness** to test for **safety** and **effectiveness**.

It is important to carry out long-term trials because sometimes the side effects may only become apparent after a significant period of time. Even if a drug becomes commonly issued, patients are still monitored to check that it is still safe. The drug can be withdrawn if the benefits no longer outweigh the risks.

Clinical trials normally compare the effects of new drugs to old ones. They have to be carefully planned to ensure the results are as accurate and reliable as possible. Patients have to agree to be part of a trial.

There are three types of trial:

1. **Open-label Trial**
    An open-label trial is where both the doctor and patient know that they are using a new treatment. This is used when the new treatment is very similar to the original, or when a drug is being compared to physical therapy.

2. **Blind Trial**
    A blind trial is where the doctor knows which treatment (i.e. the treatment or the control) the patient is receiving, but the patient does not. The idea is to remove bias – the patient may give biased information if they know which treatment is being given. An example would be using a new type of surgery on a patient.

3. **Double-blind Trial**
    A double-blind trial is where neither the patient nor the doctor administering the treatment knows whether the patient is receiving the treatment or the control. This removes the possibility of both the doctors and the patients introducing any bias.

The most rigorous trial is the double-blind trial. However, sometimes it is impossible to stop a patient or doctor from realising what treatment is being given, e.g. if the new drug has a different taste or has physical effects on the body.

Placebos

Placebos (dummy drugs containing no medication) are occasionally used in medical trials. However, they are not common practice because they create an **ethical dilemma**.

Trials involving placebos benefit society because they help to establish whether a new drug is effective or not. However, when doctors give sick patients a placebo rather than the real treatment they are offering them **false hope**. The patient believes that the pill will cure them but the doctor knows it will not.

It is also difficult to disguise a placebo. If a new drug is expected to produce certain side effects and the patient does not display them, they may work out that a placebo has been given.

For example, if a diuretic (a drug increasing urine production) is being tested, it would be easy to work out which patient had received the placebo.

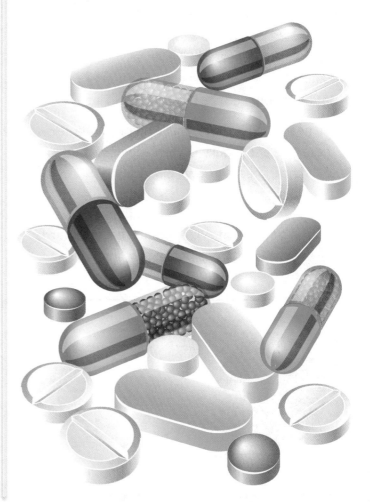

The Heart

The heart is a muscular organ in the circulatory system. It beats automatically, pumping blood around the body to provide cells with oxygen and dissolved food for **respiration**. The blood removes carbon dioxide and water as waste products.

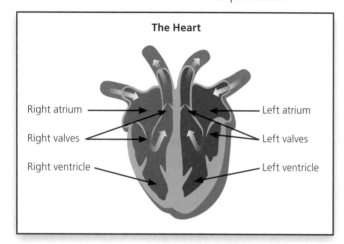

The Heart

Right atrium
Right valves
Right ventricle
Left atrium
Left valves
Left ventricle

Blood from the rest of the body enters the right atrium of the heart. It then moves into the right ventricle before being pumped to the lungs. When the oxygenated blood returns to the heart, it enters the left atrium. It then moves into the left ventricle before being pumped to the rest of the body. The heart is called a **double pump** because blood returns to it twice.

The heart itself is mainly made up of muscle cells. These cells also require oxygen and dissolved food, so the heart needs its own blood supply.

Arteries, Veins and Capillaries

Arteries carry blood away from the heart **towards** the organs. Substances cannot pass through the artery walls.

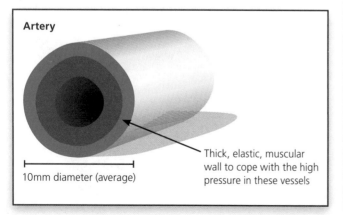

Artery

10mm diameter (average)

Thick, elastic, muscular wall to cope with the high pressure in these vessels

Veins carry blood from the organs back to the heart. Substances cannot pass through the walls of a vein.

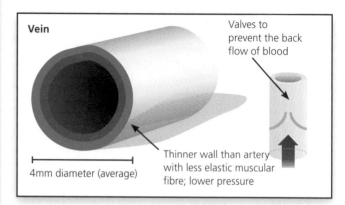

Vein

Valves to prevent the back flow of blood

4mm diameter (average)

Thinner wall than artery with less elastic muscular fibre; lower pressure

Capillaries are narrow, thin-walled vessels that allow blood to move through one cell at a time. Dissolved gases and nutrients can move out of the capillary into the surrounding cells. Waste products move back into the blood.

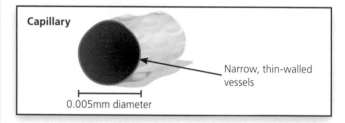

Capillary

Narrow, thin-walled vessels

0.005mm diameter

Heart Disease

Heart disease is an abnormality of the heart that can lead to a **heart attack**. It is usually caused by **lifestyle** and / or **genetic factors**, not by infection by microorganisms. Lifestyle factors that can lead to heart disease include:

* poor diet
* cigarette smoking
* misuse of drugs (e.g. alcohol, nicotine, Ecstasy, cannabis)
* stress.

Fatty deposits can build up in the blood vessels supplying the heart. This restricts the blood flow and the muscle cells do not get enough oxygen and nutrients. This can cause a heart attack.

Heart disease is more common in the UK than in non-industrialised countries. This is probably because people in the UK tend to be less active and the typical diet in the UK is high in salt and fats.

Reducing the Risk

There are precautions people can take to reduce the risk of heart disease. One of the easiest is to exercise regularly with the aim of raising the heart rate without putting it under too much stress, e.g. 20 minutes of brisk walking every day.

Health professionals can use information about a person's lifestyle, together with genetic data (family history and genetic tests), to give an indication of how likely that person is to suffer from heart disease. If the risk is high then steps, such as those listed below, can be taken to reduce the risk of heart failure:

- Do not smoke.
- Reduce salt intake in the diet.
- Maintain a healthy body weight.
- Monitor cholesterol levels (and use cholesterol-reducing drugs and foods if necessary).

Testing the Heart

The heart rate can be measured by taking the pulse. If it is too fast or slow, then it could indicate problems. Misusing drugs (such as Ecstasy, cannabis, nicotine and alcohol) can cause negative effects on health. These include altering the heart rate and blood pressure. The risk of suffering a heart attack increases.

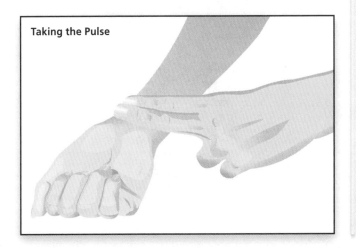

Taking the Pulse

Blood pressure is the pressure of the blood against the walls of the arteries and it results from two forces:

- **Systolic** pressure from the heart as it contracts and pumps blood into the arteries and through the circulatory system.
- **Diastolic** pressure from the force of the arteries as they resist the flow when the heart relaxes.

The pressure of the blood against the walls of the arteries can also be measured using a **sphygmomanometer**.

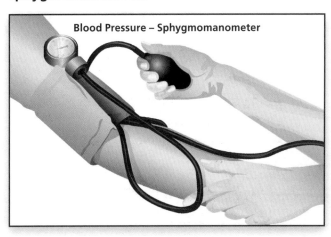

Blood Pressure – Sphygmomanometer

The sphygmomanometer gives two numbers and gives the force needed to move the metal mercury inside (hence the unit is **mmHg**):

The numbers are reported as a fraction, e.g. 118 over 76 means the systolic pressure is 118 mmHg and diastolic pressure is 76 mmHg. An average blood pressure reading is 128/80 mmHg.

A high blood pressure reading increases the chance of heart disease. 'Normal' measurements for pulse and blood pressure are always given as ranges. This is because people are not identical.

Studying Heart Disease

As heart disease is a big killer worldwide, studies continue to try to identify what factors cause it. These are called epidemiological studies and they try to identify whether a factor present in a large number of sufferers is the cause. In addition, there are more and more genetic studies taking place to identify the genes responsible for heart disease.

Homeostasis

Homeostasis is the maintenance of a constant internal environment. It is achieved by balancing bodily inputs and outputs, using the nervous system and hormones to control the process.

The body has automatic control systems which ensure that the correct, steady levels of different factors, e.g. temperature and water, are maintained. These factors are important for cells to be able to function properly.

For homeostasis to work, these control systems need to have:

- receptors (sensors) to detect changes in the environment
- processing centres to receive information and coordinate responses automatically
- effectors that produce the response.

Negative Feedback

When receptors detect that the temperature in the body has increased above a certain level, the processing centre (the brain) sends signals to the effectors (in this case sweat glands) to produce sweat to cool down the body. This process, where the steady state of the body is adjusted to reverse the change, is called **negative feedback**.

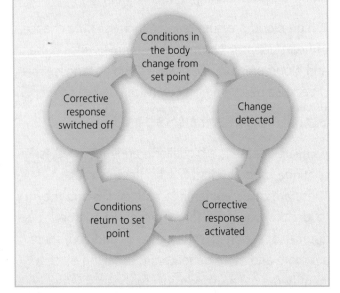

Water Balance

Water is input (gained) from drinks, food and respiration. It is output (lost) through sweating, breathing and the excretion of faeces and urine. The body has to balance these different inputs and outputs to ensure that there is enough water inside cells for cell activity to take place.

Most people have two kidneys, one situated on either side of the spine on the back wall of the abdomen. It is the job of the kidneys to control the balance of water in the body. This is achieved by adjusting the amount of urine that is excreted from the body.

The kidneys filter the blood to remove all waste (urea) and to balance levels of other chemicals (including water). The body achieves this balance through several stages:

1. Filtering small molecules from the blood to form urine (water, salt and urea).
2. Reabsorbing all the sugar for respiration.
3. Reabsorbing as much salt as the body requires.
4. Reabsorbing as much water as the body requires.
5. Excreting the remaining urine, which is stored in the bladder.

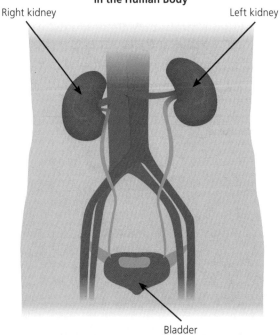

Location of the Kidneys and Bladder in the Human Body

The brain monitors water content constantly and causes the kidneys to adjust the concentration and volume of urine produced.

If the water level in the body is too low, more water is reabsorbed by the kidneys. If the water level is too high, the urine becomes more dilute and watery.

The amount of water that needs to be reabsorbed depends on a number of factors:

- **External temperature**

  High → concentrated urine

  Low → dilute urine

- **Level of exercise**

  High → concentrated urine

  Low → dilute urine

- **Fluid intake**

  High → dilute urine

  Low → concentrated urine

- **Salt intake**

  High → dilute urine

  Low → concentrated urine

The concentration of urine is controlled by a hormone called **ADH**, which is released into the bloodstream by the **pituitary gland**.

When the level of water in the blood is too low, ADH is released and this causes concentrated urine to be produced. This is because the hormone causes the kidney to become more permeable, allowing water to be reabsorbed.

When the level of water in the blood is too high, ADH is not released. The kidney becomes less permeable and this causes dilute, watery urine to be produced.

This is another example of negative feedback.

Effect of Alcohol

Alcohol is a drug that causes the production of a greater volume of dilute, watery urine. This can lead to dehydration and other adverse effects on health. The symptoms of dehydration include headaches, which, unsurprisingly, are also a symptom of hangovers.

The reason why alcohol leads to the production of greater volumes of urine is because ADH is suppressed.

Effect of Ecstasy

Ecstasy is a drug that makes people feel euphoric. It can increase the heart rate and blood pressure, which can cause long-term health problems, including increasing the chance of having a heart attack.

The drug also interferes with the brain, causing errors in monitoring the water content of the body. Although someone taking Ecstasy may get hot and drink a lot of water, the brain fails to send messages to the kidneys to get rid of the extra water.

The urine that is produced is concentrated when it should be dilute. As the water cannot escape, it causes cells to swell up. Cells in the brain get squashed against the skull and die. This may result in death.

The reason why Ecstasy leads to the retention of greater volumes of water is because it causes increased ADH production.

Effects of Cannabis and Nicotine

Cannabis and nicotine are drugs that are smoked. They have the effect of increasing the heart rate. Cannabis reduces blood pressure whilst nicotine increases it. These effects can cause long-term heart problems and increase the risk of a heart attack.

All life on Earth is adapted for living in particular habitats. This module looks at:

- how living things are adapted to the environment
- how nutrients such as carbon and nitrogen are cycled
- how life evolved
- the evidence for evolution.

Adaptations

A group of organisms that can breed together and produce fertile offspring is called a **species**.

The individuals in a species are adapted to living in their environment. This means that their features work best with the environment that they live in. For example, polar bears are adapted to living in the Arctic and tigers are adapted to the jungle.

All living things rely on their environment and other species to survive. A **habitat** is the term used to describe the environment in which an animal or plant lives.

The adaptation of living organisms to their environment enables the organisms to live longer. If an organism survives to sexual maturity, it is more likely to pass on its genes, including the genes coding for its adaptations, to the next generation. Adaptations therefore can lead to increased chances of survival.

When the same **resources** (e.g. shelter, food, water, light availability, etc.) are needed by different organisms in the same habitat then there is **competition**. The organisms that are most successful at competing survive and pass on those genes that code for the adaptations.

Cactus

The cactus is a plant native to America that is supremely adapted for living in conditions of high temperature and low rainfall.

A cactus plant has the following adaptations that enable it to survive in the environment that it lives in.

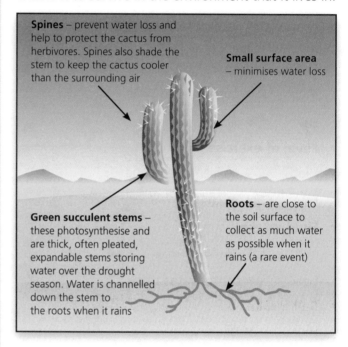

Spines – prevent water loss and help to protect the cactus from herbivores. Spines also shade the stem to keep the cactus cooler than the surrounding air

Small surface area – minimises water loss

Green succulent stems – these photosynthesise and are thick, often pleated, expandable stems storing water over the drought season. Water is channelled down the stem to the roots when it rains

Roots – are close to the soil surface to collect as much water as possible when it rains (a rare event)

Cacti are typically found in hot, dry climates. Their adaptations enable them to survive and pass on their successful genes to the next generation.

The table shows the number of cacti per km^2 in different areas of North America.

US State	Average August Temperature (°C)	Number of Cacti per km^2
Texas	23.6	100
Mexico	24	120
California	21	70
Alaska	16	0

Looking at the table, we could explain that cacti are found in hot climates because their adaptations (spines, green succulent stems, roots close to the surface and small surface area) enable them to survive where other plants cannot.

They do not do well in colder climates, such as in Alaska, because their adaptations are not suited to colder climates.

Food Chains and Food Webs

Food chains show the direction of energy and material transfer between organisms.

Food webs show how all the food chains in a given habitat are interrelated. In practice, these can be very complex because many animals have varied diets. All the organisms in a food web are dependent on other parts of the web.

 This is called **interdependence**.

A Food Web

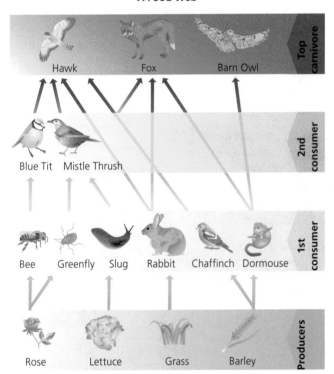

Changes to the environment can alter numbers in the food web. For example, a small decrease in the amount of rain could reduce the amount of lettuce available and cause reductions in the number of slugs.

If the changes are too great for the natural variation within a population to cope with, then organisms will die out before they can reproduce. The population will decline and eventually become extinct (disappear completely).

The same process happens if a new species that is better at competing for resources (a competitor) is introduced. A new predator or a disease organism for a species can have the same effect.

If a species (animal, plant or microorganism) becomes extinct, this can affect other parts of the food web and cause further extinctions.

Energy Flow Through Food Chains

Energy from the Sun enters the food chain when green plants absorb light in order to photosynthesise. When animals eat the plants, the energy passes up the food chain from one organism to another.

Energy in a food chain flows in one direction:

1. A small proportion of the Sun's light energy transfers to an **autotroph** (a plant), which captures the energy, carries out photosynthesis and stores the energy in chemicals (such as cellulose) in its cells.
2. A **herbivore heterotroph** then eats the autotroph. Some of the energy stored in the plant is transferred to the herbivore and stored in its cells.
3. A **carnivore heterotroph** then eats the herbivore heterotroph. Some of the energy stored in the herbivore is transferred to the carnivore and stored in its cells.

Energy Flow in a Food Chain

At each stage of the food chain, a large proportion of the energy is:
- lost to the environment as heat
- excreted as waste products
- trapped in indigestible material such as bones, cellulose and fur.

This means that as the food chain moves from autotrophs to heterotrophs, there is less energy available at each **trophic level** (i.e. each stage of the food chain). Therefore there is a limit to the number of levels in each food chain. The limit is usually four or five levels.

Decay Organisms

Energy is also transferred by decay organisms that break down organisms after they die.

There are two types of decay organism:

- **Decomposers**, such as bacteria and fungi, break down the dead material and use the energy stored inside.

- **Detritivores** include animals such as earthworms and woodlice. These consume the detritus (dead plants or animals and their waste), breaking it down into smaller particles that other detritivores and decomposers can use.

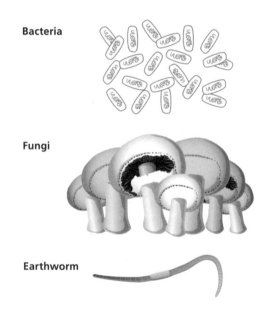

Bacteria

Fungi

Earthworm

Calculating Energy Efficiency

The percentage of energy efficiency can be calculated using the following formula:

$$\text{Percentage of energy successfully transferred} = \frac{\text{Amount used}}{\text{Amount potentially available}} \times 100$$

Example

The arrow diagram to the right shows the feeding relationship between a green plant, a caterpillar and a bird.

Calculate how efficient the energy transfer is for the caterpillar feeding on the plant.

$$\text{Percentage of energy successfully transferred} = \frac{\text{Amount used}}{\text{Amount potentially available}} \times 100$$
$$= \frac{80}{800} \times 100$$
$$= 10\%$$

On average, only 10% of the energy from the Sun ends up being stored as plant tissue. If an animal eats a plant, 10% of the energy gained is used to build up biomass. The rest of the energy is used by the animal to respire, move and keep warm.

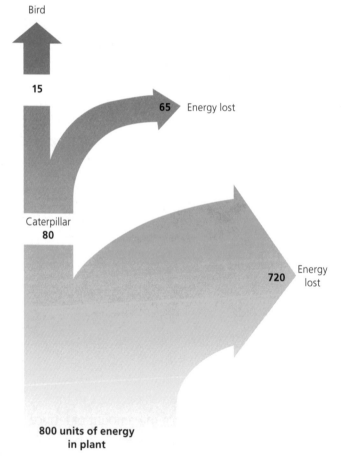

Bird

15

65 Energy lost

Caterpillar
80

720 Energy lost

800 units of energy in plant

Carbon

Carbon is a vital element for living things. It is used in all organic molecules, including sugars, proteins and amino acids. Life on Earth is very much carbon-based.

Carbon is recycled through the environment so that it is available for life processes. This can be seen in the **carbon cycle**:

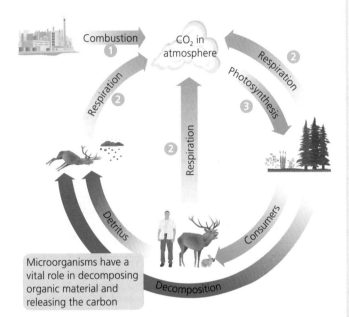

Microorganisms have a vital role in decomposing organic material and releasing the carbon

① Combustion

$$Fuel + Oxygen \longrightarrow Carbon\ dioxide + Water + Energy\ released$$

② Respiration

$$Glucose + Oxygen \longrightarrow Carbon\ dioxide + Water + Energy\ released$$

③ Photosynthesis

$$Carbon\ dioxide + Water \xrightarrow{Light\ energy} Glucose + Oxygen$$

Nitrogen

Another element vital for life is nitrogen. Nitrogen gas makes up 79% of the Earth's atmosphere. Nitrogen molecules have a triple covalent bond between the atoms of nitrogen and this makes it impossible for most organisms to break the bond and use the nitrogen.

Nitrogen is 'fixed' into a form that plants and animals can use in two ways: lightning strikes and the action of specialised bacteria.

Nitrogen, like carbon, has to be recycled to ensure that it is available for life processes. This can be seen in the **nitrogen cycle**:

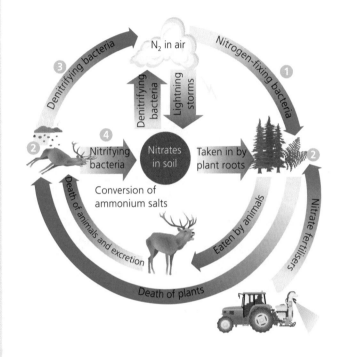

HT ① Nitrogen fixation – carried out by bacteria such as *Azotobacter sp.* and blue green algae such as *Anabaena variabilis*

② Conversion of nitrogen compounds into proteins inside plants and animals

③ Denitrification – this is where, in the absence of oxygen, denitrifying bacteria such as *Thiobacillus denitrificans* convert nitrates in dead plants and animal remains back into nitrogen gas. This completes the cycle, releasing nitrogen back into the atmosphere.

④ Microorganisms convert ammonia to nitrites and then to nitrates via the process of nitrification.

The nitrogen that is now in compounds in plant material is passed through the food chain through animals eating the plants.

Measuring Environmental Changes

Biologists can measure changes in the environment by using indicators. These may be non-living or living.

Non-living Indicators

Nitrate levels can be measured using test kits with chemicals that change colour. The chemicals can then be matched against a chart indicating the amount of nitrate present in the sample.

Temperature can be measured using a thermometer, or a data-logger, which is more accurate and reliable.

Carbon dioxide levels can be measured using data-loggers.

Biological Indicators

Changes in the environment affect living organisms. Biologists can use changes in the pattern of where a species is living to determine how climate change or other environmental changes, such as road-building projects, are affecting the species.

In the oceans, **phytoplankton** (microscopic plants) are useful for detecting the effect of temperature changes and for detecting changes in the food web.

Lichens grow very slowly and are susceptible to atmospheric pollutants and acid rain. A decline in their number can indicate pollution.

River organisms, such as the larvae of mayfly (called **mayfly nymphs**), can be used to indicate the quality of the water. Mayfly nymphs can only live in clean river water with enough oxygen. If a river has mayfly nymphs, then pollution levels will be low.

River pollution typically causes mass growth of bacteria, which then use up all the oxygen. This causes fish to die and indicator species such as **bloodworm** and **rat-tailed maggots** to grow instead, as these are pollution tolerant and can survive in low oxygen concentrations.

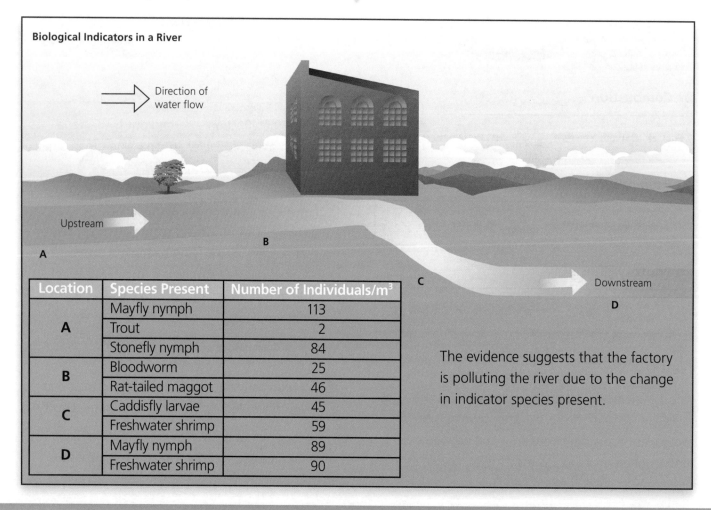

Biological Indicators in a River

Direction of water flow

Upstream

A

B

C

Downstream

D

Location	Species Present	Number of Individuals/m³
A	Mayfly nymph	113
	Trout	2
	Stonefly nymph	84
B	Bloodworm	25
	Rat-tailed maggot	46
C	Caddisfly larvae	45
	Freshwater shrimp	59
D	Mayfly nymph	89
	Freshwater shrimp	90

The evidence suggests that the factory is polluting the river due to the change in indicator species present.

How Life Evolved

Life on Earth began around 3500 million years ago. **All** life on Earth, including all life that is now extinct, evolved from very simple living things. So all organisms share a common ancestor.

The naturalist Charles Darwin thought creatively about the problem of how organisms are related. Creative thinking is a skill scientists need when developing explanations. To show the linked ancestors of different species, he drew a tree of life to illustrate his thinking:

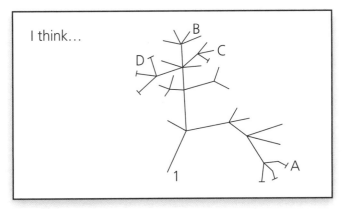

The evidence of the fact that organisms share a common ancestor comes from:

1 The Fossil Record

Fossils are the remains of plants or animals from thousands of years ago that are found in rock. Fossils indicate the history of species and can show the evolutionary changes in organisms over millions of years. Fossils can be formed from the hard parts of animals that did not decay easily, or from parts of plants or animals that did not decay

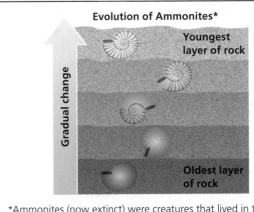

Evolution of Ammonites*

Gradual change

Youngest layer of rock

Oldest layer of rock

*Ammonites (now extinct) were creatures that lived in the sea millions of years ago.

because one or more of the conditions needed for decay were absent, e.g. oxygen, moisture and temperature. The remains of the organisms are then buried until they are rediscovered.

2 Genomics

Similarities and differences in DNA can lead to the relationships being worked out between all life on Earth. Analysing the DNA of both living and fossilised specimens shows that there are similarities as well as differences. This can be used to chart the family tree of all life on Earth. The more shared genes organisms have, the more closely related they are.

Mapping the differences and similarities enables the family relationships between organisms to be measured. For example, the diagram below (a **phylogenetic tree**) shows the relationship between the different genes for mouse taste receptors (blue) and the taste receptors in humans (gold). The red markers indicate where the taste receptor is the same in mice and humans. We share 11 taste receptors with mice and approximately 85% of all our genes.

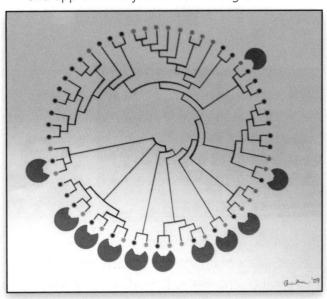

All organisms have DNA. Organisms in a species differ from one another, e.g. offspring look different to their parents. These changes are due to the environment and genetic differences in the DNA. Only the **changes** in DNA (mutations) can be passed on to the offspring.

Gene Mutation

Genetic variation is caused when changes called **mutations** take place in the genes. Mutations cause different proteins to be produced and this changes the function of the gene. Sometimes a mutation causes a gene to be copied twice, which means that more variation can occur.

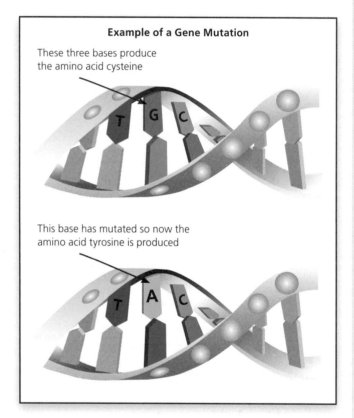

Example of a Gene Mutation

These three bases produce the amino acid cysteine

This base has mutated so now the amino acid tyrosine is produced

If the mutations occur in the cells producing eggs in the ovaries and sperm in the testes, then the mutated genes may be passed on to the offspring. Most of the time the mutations have no significant effect. Sometimes the mutation causes new characteristics.

Natural Selection

The genetic variation between individuals in a species means that those with characteristics that improve their chances of survival in their physical environment are more likely to live to adulthood.

When these individuals reproduce, they pass on the beneficial characteristics to their offspring. Individuals with characteristics poorly-fitting to their environment are less likely to survive.

As a result, the number of individuals in a population with beneficial traits increases while the number of individuals with non-beneficial traits decreases.

The selective agent is the organism's environment (e.g. availability of water, availability of food, space and predation). If the selective agent changes, then what may have been a non-beneficial trait may become beneficial, giving those organisms with the trait more success.

Selective Breeding

Selective breeding is where animals and plants with certain traits are deliberately mated together (crossed) to produce offspring with certain desirable characteristics.

① Creating New Varieties of Organisms

Dalmation Dogs

Choose the spottiest two to breed...

... and then the spottiest of their offspring...

... to eventually get Dalmations.

② Increasing the Yield of Animals and Plants

Some types of cattle have been bred to produce high yields of milk, or milk with a low fat content.

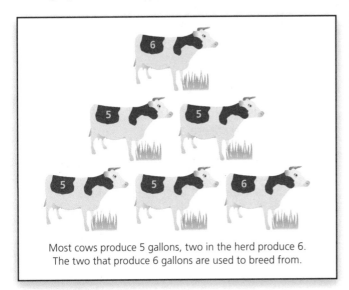

Most cows produce 5 gallons, two in the herd produce 6. The two that produce 6 gallons are used to breed from.

Improved crops can be obtained through selective breeding programmes, although this happens over a long period of time.

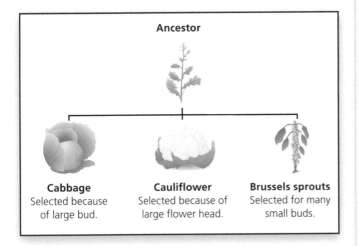

Ancestor

Cabbage	**Cauliflower**	**Brussels sprouts**
Selected because of large bud.	Selected because of large flower head.	Selected for many small buds.

Natural Selection and Selective Breeding – Differences and Similarities

Differences

With selective breeding, it is humans who choose the desirable traits. With natural selection, it is the environment that determines the desirable traits.

Similarities

Both natural selection and selective breeding act on the genetic variation within a population.

Peppered Moths

Peppered moths are usually pale and speckled in colour. This means that they are well-camouflaged against the bark of silver birch trees.

Within the population is a trait that causes the moth to be completely black. The black moths are at a disadvantage because predators can see them more easily against the tree. When pollution from factories covered the trees with black soot, the situation was reversed.

The Clean Air Act in the 1950s caused the amount of pollution to decrease drastically:

Numbers of Peppered Moths Caught in a Day		
Data from 1900		
	Black	**Speckled**
Industrial	135	29
Countryside	5	180
Data from 2000		
	Black	**Speckled**
Industrial	9	160
Countryside	7	171

Conclusion: the coloration of the moth was selected for by the environmental conditions.

Peppered Moth

Black Peppered Moth

Evolution

The combined effects of the following can lead to the formation of new species:

- Mutations
- Natural selection
- Environmental changes
- Isolation – where individuals from one population are isolated from other populations so that they cannot meet to breed.

This process is called **evolution**.

Jean-Baptiste Lamarck was a scientist who proposed that the environment changed an organism. The organism then passed on the characteristic to their offspring, e.g. moles lived in the dark, so they lost their eyes as a result. This is called **evolution through inheritance of acquired characteristics**.

Charles Darwin devised a better explanation following many years of thought and collecting evidence. By collecting data, Darwin made the connection between varieties, competition, the survival of the fittest and the passing on of desirable characteristics to the next generation.

This is an example of how the science community works. Darwin had come up with a scientific explanation that was better than previous explanations, such as Lamarck's. It fitted the available evidence at the time, and it fits with our modern understanding of genetics today. There was, however, no evidence or scientific mechanism for Lamarck's inheritance of acquired characteristics. The scientific community, having repeated Darwin's experiments and peer reviewed his work, accepts Darwin's explanation for evolution over Lamarck's.

Today it is accepted that **all** life on Earth arose through the process of **evolution through natural selection**.

Biodiversity

Biodiversity refers to the variety of life on Earth. It can be measured in a number of ways:

- The number of different species present in an area
- The range of different types of organism (e.g. plants, animals and microorganisms)
- The genetic variation within species (how many different alleles there are).

The greater these are, the greater the biodiversity.

Biodiversity is important because it enables ecosystems and the species within them to survive natural disasters. For humans it is vitally important because we need to exploit crops to feed a growing population. Plants are also the source of compounds that are effective against disease and genetic disorders. These could potentially be used as medicines. For example, quinine, a drug effective against malaria, was discovered and extracted from the quinchona tree in South America.

If ecosystems are destroyed and biodiversity is reduced, then drugs and food substances may be lost for future generations.

Classification

All life on Earth can be divided into groups based on similarities and differences in their physical features (e.g. flowers in flowering plants, the skeleton in vertebrates) and in their DNA.

The groups start off large with high numbers of organisms with a few features in common (e.g. **Kingdoms** such as plants, animals) and reduce in size as they are sub-divided into much smaller groups containing organisms with similar features (e.g. **species** such as *Homo sapiens*).

Classifying both living and fossil organisms can help to make sense of the enormous diversity of living things on Earth as well as to show the evolutionary relationships between organisms.

Extinction

Throughout the history of Earth, species of animals and plants have become extinct (i.e they do not exist anywhere on the planet), e.g. the dodo.

The rate of extinctions on Earth has been increasing. This is likely to be due to human activity. Humans can cause extinctions directly or indirectly:

Direct Causes
• Excessive hunting of animals
• Removal of habitats to extract resources (e.g. timber from rainforests) or for building

Indirect Causes
• Introducing predators to new locations (e.g. when colonising Australia)
• Activities causing global warming – the activities of humans indirectly change the environmental conditions
• Trying to eradicate a pest – this can have a knock-on effect on other members of a food web

Sustainability

Sustainability is about meeting the needs of people today without damaging the Earth for future generations.

Farming

In the past, farmers used to grow a variety of different crops on a smallholding. Hedgerows would separate the different parts of the farm. In the 20th century, techniques changed. Giant fields made from many earlier fields joined together were planted with **monocultures** (a single variety of a crop).

This change to intensive farming has advantages because it maximises the amount of food available for the population. However, it reduces biodiversity. Hedgerows have decreased in number and the food chains that rely on them have also decreased. Intensive farming is not sustainable because it does not maintain biodiversity.

However, the solution is not to revert to old-fashioned methods of farming. To ensure all people can have food, scientists need to work out new ways of optimising intensive farming while maximising sustainability.

Improving Sustainability

Virtually all products used in the industrialised world rely on oil and the products made from it. To improve sustainability, alternatives to oil need to be found.

Packaging is used to attract the attention of consumers, as well as providing a way of keeping the product safe. Packaging is often made from oil-based plastics.

Manufacturers have to consider:
- what **materials** should be used
- how much **energy** is needed in the manufacturing process for a given packaging material
- how much **pollution** will be produced as a result of manufacturing packaging.

For example, crude oil is a fossil fuel that takes millions of years to form and the plastic that is made from oil will not biodegrade. Using oil is therefore unsustainable, as it cannot be remade.

Sustainable alternatives would include using a packet made from paper or from cellulose-based plastics. The type of plants that produced the material can then be re-planted.

Packaging ends up being thrown away. The rubbish is taken to **landfill** sites. Even if the packaging is biodegradable, there will still be slow decomposition as there is a lack of oxygen in landfill.

Manufacturing and transporting packaging also use up energy and produce pollution. It would therefore be far more preferable to reduce the amount of packaging used for products.

For example, Easter eggs are sold in bright boxes with a plastic insert (made from oil) to display the egg, which is encased in foil. Some companies are now selling eco-friendly Easter eggs. These consist of just the egg wrapped in paper or foil. This reduces the waste, pollution and energy costs for what is a short-lived product.

Exam Practice Questions

1 This question is about alleles.

(a) If the gene for having dimples was dominant, the allele would be written as a D. Which of the following individuals has dimples? Put ticks (✓) in the boxes next to the **two** correct answers. **[1]**

dd ☐   DD ☐   Dd ☐

(b) Complete the genetic cross between a male (with Bb alleles) and a female with bb alleles, using a Punnett square. **[2]**

♂

	....	....
....	....	....
♀ | .... | .... | .... |

2 This question is about genetic disorders.

(a) Which of the following are symptoms of cystic fibrosis?
Put ticks (✓) in the boxes next to the **three** correct answers. **[3]**

Runny nose ☐   Nausea ☐   Ear ache ☐   Breathing difficulties ☐

Thick mucus in lungs ☐   Forgetfulness ☐   Chest infections ☐   Swollen glands ☐

(b)

Every pregnancy should be tested for Huntington's disease and the fetus aborted if the condition is present.

At least people live to middle age before the symptoms of Huntington's disease start to develop.

Linda   **Jamie**

Write a response to Linda explaining why her idea is potentially unethical. **[6]**

✏ *The quality of written communication will be assessed in your answer to this question.*

3 What are the two ways that humans have directly caused an organism's extinction? **[2]**

4 The following statements describe the different stages in the vaccination process. They are in the wrong order. Fill in the empty boxes to put the stages in the correct order. **[2]**

A The markers on the surface of the microorganism trigger the production of specific antibodies by white blood cells.

B The white blood cells capable of fighting the microorganism remain in the bloodstream.

C The modified microorganism is injected into the body.

D A harmful microorganism is modified so that it is incapable of multiplying.

E The microorganism is destroyed before it causes harm.

Start | | | | | |

5 Explain the disadvantages of making genetic testing compulsory. **[6]**

✐ *The quality of written communication will be assessed in your answer to this question.*

6 Label the heart. **[3]**

HT 7 Draw straight lines to join each term to the **best** available explanation. **[4]**

Term	Explanation
Genotype	A version of a gene
Phenotype	Possessing two of the same alleles
Allele	The characteristics expressed in the environment
Heterozygous	Possessing one of each allele type
Homozygous	The alleles present for a gene in an individual
	A gene

8 Which of the following is the **best** explanation for why at least 95% of the population needs to be vaccinated? Put a tick (✓) in the box next to the correct answer. **[1]**

This level stops the disease from spreading between vaccinated people. ☐

People with the disease cannot pass it on. ☐

Enough of the population is vaccinated to avoid an epidemic. ☐

95% ensures the maximum profit for the medical company. ☐

9 A caterpillar takes in 60kJ of energy. It loses 54kJ of energy through movement, other body processes and waste. The caterpillar is then eaten by a blackbird. What percentage of energy entering the caterpillar is transferred to the blackbird? **[2]**

Many different processes take place inside every cell of a living organism. This module looks at:

- the provision of molecules for energy and carbon skeletons via photosynthesis
- respiration under aerobic and anaerobic conditions
- enzymes and how they work
- the processes by which molecules enter and leave cells
- how to carry out sampling in fieldwork.

Chemical Reactions in Living Things

All living things are made from basic units. These are called **cells**.

Chemical reactions take place inside cells to enable them to function. All chemical reactions need energy to take place. The energy comes from the process of **respiration** – the release of energy from food.

The energy in food ultimately comes from the Sun. Plants (and some microorganisms, such as phytoplankton) at the start of food chains capture light energy by the process of photosynthesis, making food molecules. Energy is then transferred through food chains.

Photosynthesis

Photosynthesis can be summarised by saying that carbon dioxide and water are combined to produce sugar and oxygen in the presence of light and chlorophyll.

This sounds simple enough but in reality it just tells you the reactants needed at the start and the products created at the end. In reality photosynthesis is a series of reactions, some taking place in the light and some in the dark.

The sugar produced by photosynthesis is used as a carbon skeleton to build more organic molecules. All life is built from carbon skeletons in this way.

Respiration

Respiration can be summarised by saying that sugar and oxygen are combined to produce carbon dioxide and water, releasing energy in the process.

Respiration can be thought of as slow combustion. The sugar is the fuel.

As with photosynthesis, the summary above just tells you the reactants needed at the start and the products created at the end. Respiration is, in fact, a series of reactions releasing energy from large food molecules in all living cells.

Enzymes

Enzymes are organic catalysts. They are protein molecules that speed up the rate of chemical reactions in living organisms.

Enzymes are a type of protein coded for by instructions carried in genes.

The Lock and Key Model

Only a molecule with the correct shape can fit into an enzyme. This is a bit like a key (i.e. the molecule, called a substrate) fitting into the lock (i.e. the enzyme). Once the enzyme and molecule are linked the reaction takes place, the products are released and the process is able to start again.

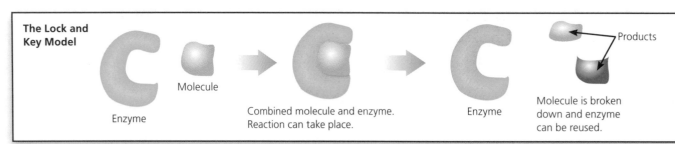

The Lock and Key Model

Enzyme

Molecule

Combined molecule and enzyme. Reaction can take place.

Enzyme

Products

Molecule is broken down and enzyme can be reused.

Enzymes and Temperature

For enzymes to work to their optimum they need a specific and constant temperature. The graph below shows the effect of temperature on enzyme activity.

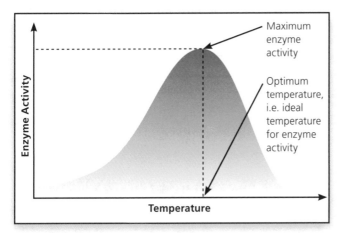

Different enzymes have different optimum working temperatures. For example, in the human body enzymes work best at about 37°C. Below this temperature their rate of action slows down while above 40°C they permanently stop working.

> **HT** The enzyme activity at different temperatures is a balance between the increase in rate of reaction as the temperature increases and the changes to the active site at higher temperatures.

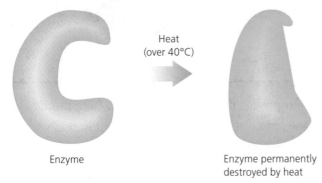

Enzymes also have an optimum pH. For example, salivary amylase (an enzyme that breaks down starch in the mouth) works best at pH 6.8 (slightly acidic). Meanwhile, pepsin (an enzyme that breaks down protein in the stomach) works best at pH 1.5 (very acidic).

Enzymes can speed up catabolic reactions (where molecules are broken down) as well as anabolic reactions (where new molecules are built up).

HT Denaturing

The biological name for the process of permanent change in an enzyme's shape is **denaturing**. The enzyme becomes denatured.

At low temperatures, a small rise in temperature causes an increase in the rate of reaction.

The enzyme activity increases until the optimum temperature is reached. After this, the enzymes will be denatured – the active site (see below) will now be a different shape and the substrate will no longer fit.

A good analogy is a beef burger, made of protein. Before cooking, the burger is a particular shape. After cooking at high temperature, its shape is irreversibly changed.

The Active Site

The place where the substrate fits to the enzyme is called the **active site**. Each enzyme has a specific active site – only certain substrates will fit with it.

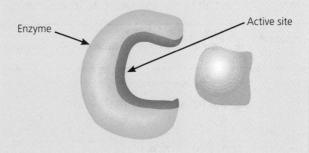

When subjected to high temperature, the shape of the active site **changes irreversibly**. This means molecules can no longer fit and the reaction stops. The enzyme is denatured.

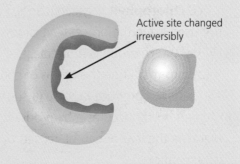

pH Changes

The active site is influenced by pH. Changes in pH in the enzyme's environment can make and break intra- and intermolecular bonds in the enzyme. This changes the shape of the active site and its effectiveness.

The graph below shows how changes in pH levels affect enzyme activity.

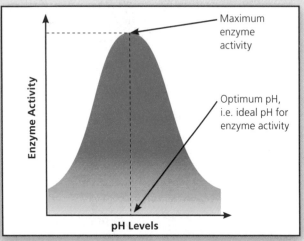

Photosynthesis

Photosynthesis can be written as a word equation:

$$\text{Carbon dioxide} + \text{Water} \xrightarrow{\text{Light energy}} \text{Glucose} + \text{Oxygen}$$

HT The equation can also be written as a chemical formula:

$$6CO_2 + 6H_2O \xrightarrow{\text{Light energy}} C_6H_{12}O_6 + 6O_2$$

There are three stages in photosynthesis:

1. Light energy is absorbed by a green chemical called **chlorophyll** in green plants. Chlorophyll is not used up in the process – it is not a reactant.
2. Within the chlorophyll molecule, the light energy is used to rearrange the atoms of carbon dioxide and sugar to produce glucose (a sugar).
3. Oxygen is produced as a by-product. It exits the plant via the leaf or can be reused by the plant in respiration.

Glucose in Plants

Glucose is effectively a carbon skeleton that can be used to build many other molecules in living things. For example, cellulose is a structural carbohydrate that makes cell walls in plants. It is very strong and helps to provide support. Cellulose is made of repeating molecules of a certain form of glucose.

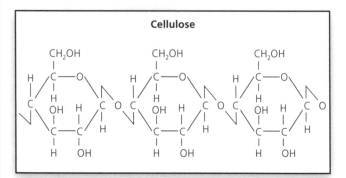

Cellulose

Protein is made up of different amino acid molecules, which themselves were made from glucose. They are vital for growth and repair.

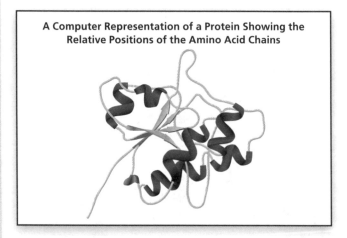

A Computer Representation of a Protein Showing the Relative Positions of the Amino Acid Chains

Chlorophyll is the organic pigment where the first stages of photosynthesis take place. It is also produced using a skeleton based on carbon.

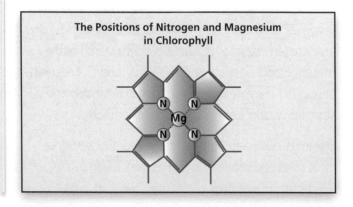

The Positions of Nitrogen and Magnesium in Chlorophyll

Storing Glucose

Glucose is converted into another carbohydrate suitable for storage, called **starch**. Starch is a long chain of glucose units that can be packed together very efficiently.

Glucose is used for energy, which is released by respiration.

Plant Cell Structure

The diagram below shows a plant cell:

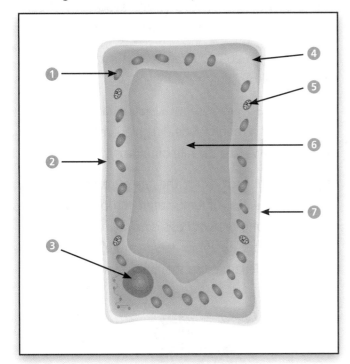

1. Chloroplasts – contain chlorophyll and the enzymes for some of the reactions of photosynthesis. They are only present in green parts of the plant
2. Cell membrane – allows dissolved gases and water to enter and leave the cell freely while acting as a barrier to other, larger chemicals
3. Nucleus – contains DNA that carries the genetic code for making enzymes and other proteins
4. Cytoplasm – where enzymes and other proteins are made
5. Mitochondria – where respiration occurs
6. Vacuole – used by the cell to store waste materials and to regulate water levels
7. Cell wall – provides support for the cell

Plant Transport

Plants need other chemicals in addition to glucose. The roots take up minerals from the soil in solution. Nitrogen, in the form of nitrates, is absorbed and used by the plant cells to make new proteins.

Substances move through cells via the process of **diffusion**. Diffusion is the overall movement of a substance from a region where it is in high concentration to an area where it is in lower concentration.

Diffusion is a passive process as it does not need an energy input to happen. For example, if the concentration of carbon dioxide in a plant cell is lower than the concentration outside, then carbon dioxide will diffuse into the cell.

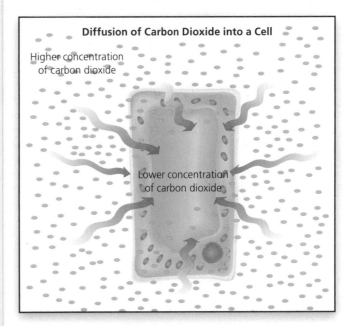

Diffusion of Carbon Dioxide into a Cell

Higher concentration of carbon dioxide

Lower concentration of carbon dioxide

Diffusion is the main method by which gases enter and leave the plant. Gases such as carbon dioxide and oxygen exchange between the leaf and the surrounding air through small holes on the underside of a leaf, called **stomata**.

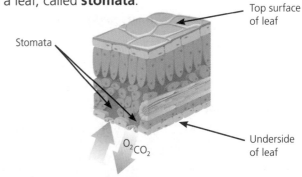

Top surface of leaf

Stomata

Underside of leaf

$O_2 \ CO_2$

Osmosis

Osmosis is a specific type of diffusion. It is the overall movement of water from a dilute (i.e. with a high water to solute ratio) to a more concentrated solution (i.e. with a low water to solute ratio) through a **partially permeable** membrane.

A partially permeable membrane allows water molecules through, but not solute molecules because they are too large.

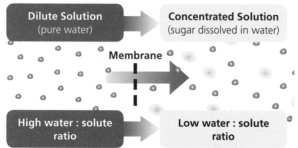

Dilute Solution (pure water)	Concentrated Solution (sugar dissolved in water)

Membrane

High water : solute ratio	Low water : solute ratio

The effect of osmosis is to gradually dilute the concentrated solution. This is what happens at root hair cells, where water moves from the soil into the cells by osmosis due to a concentration gradient.

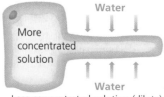

Less concentrated solution (dilute)

Water

More concentrated solution

Water

Less concentrated solution (dilute)

HT Active Transport

Active transport is the overall movement of a chemical substance against a concentration gradient (i.e. from where the substance is in low concentration to where it is in a higher concentration). This requires energy, which is provided by respiration.

An example of active transport is found in plant roots. Plants require nitrates. The nitrate concentration inside the plant is naturally higher than the outside. The plant cells use active transport to bring the nitrate from where it is in low concentration to where it is at a higher concentration in the root cell.

Limitations to Photosynthesis

There are several factors that can interact to limit the rate of photosynthesis:

- **Temperature** – too low and photosynthesis stops until the temperature rises again. Too hot and the enzymes stop working permanently.
- **Carbon dioxide concentration** – as carbon dioxide concentration increases, so does the rate of photosynthesis.
- **Light intensity** – light is needed for photosynthesis. The greater the availability of light, the quicker photosynthesis will take place.

You need to be able to interpret data on limiting factors for photosynthesis.

The graphs below each show a limiting factor (temperature, carbon dioxide or light intensity). The initial stage ❶ shows that the rate of photosynthesis increases.

At stage ❷ of each graph, the rate of reaction stays the same or, in the case of temperature, drops. This indicates that the rate of photosynthesis is now limited by one of the other factors (except for temperature, where enzymes are denatured so that photosynthesis stops).

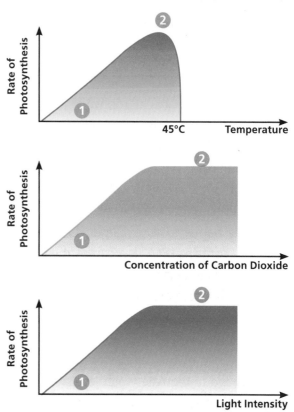

Fieldwork

To identify the effect of light on plants, biologists have to carry out fieldwork. This involves using a variety of techniques to measure the amount of available light and to see how this has affected the growth of plants.

A **light meter** can measure the amount of light that is hitting a leaf. The amount of light is measured in units of lux. Data-loggers can be fitted with a light meter and readings taken over a period of time.

It is impractical, if not impossible, to count all the plants growing in a given area. Instead, biologists sample a proportion of the available land to give an accurate estimate of plant cover. To determine this, biologists use **quadrats**. A quadrat is a square shape, often divided up into smaller squares. The quadrat is placed randomly on sections of the area in question and the plants that fall inside the area of the quadrat are counted.

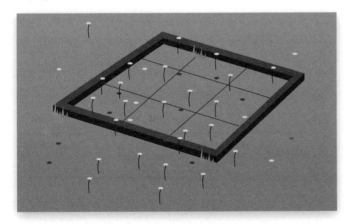

Alternatively, the area covered by the leaves of the plants can be counted. Sometimes a scale is used to indicate relative abundance, e.g. VC, C, UC, R, VR (very common, common, uncommon, rare, very rare).

It is vital to use a **key** to ensure that plants are correctly identified. A key enables the rapid identification of plants and animals by asking questions such as 'does the plant have parallel veins in its leaves?'

The answers rapidly lead to an idea of what the plant or animal is.

Example of a Key

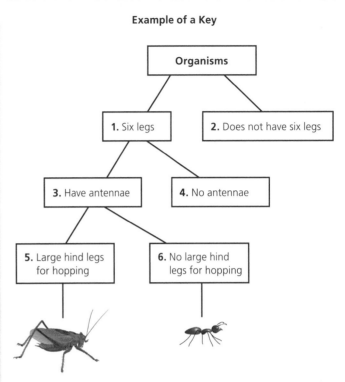

Sometimes it is preferable to measure the changes in plant life along a straight line, e.g. to see the succession of plants from a forest to the sea. In this case, a transect may be taken. A line is drawn and the quadrat is placed at set intervals along the line and the plants counted as before. This gives a picture of the changes in plant life over the line of the transect. In the example below, three fixed area plots are also investigated using randomised quadrats.

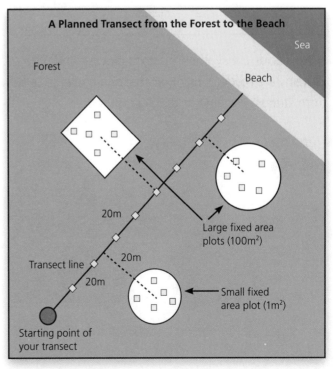

Respiration

All living organisms require energy for chemical reactions inside cells. This energy is released through the process of respiration. The energy is used for:

*   movement
*   synthesising (making) larger molecules
 *   active transport.

Synthesis of Large Molecules

Larger molecules, such as starch and cellulose, are synthesised from smaller molecules, such as glucose in plant cells. This involves joining the glucose molecules (monomers) together to form a polymer (made of many units).

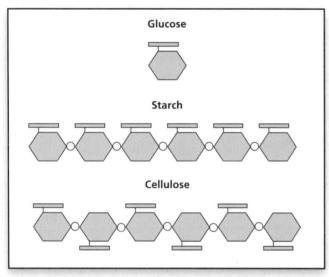

Amino acids are synthesised from glucose and nitrates. Proteins are made in plant, animal and bacterial cells from strings of amino acids joined together.

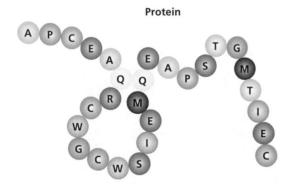

Aerobic Respiration

Aerobic respiration releases energy through the breakdown of glucose molecules, by combining them with oxygen inside living cells. The majority of animal and plant cells, and some microorganisms, respire aerobically.

The word equation for aerobic respiration is:

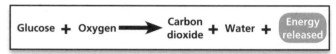

The chemical formula for aerobic respiration is:

$$C_6H_{12}O_6 + 6O_2 \longrightarrow 6CO_2 + 6H_2O$$

Anaerobic Respiration

Anaerobic respiration takes place in animal cells (e.g. in humans during vigorous exercise), plant cells (e.g. in plant roots in waterlogged soil) and in some microbial cells (e.g. bacteria in puncture wounds) when there is no oxygen or oxygen supply is low.

The equation for anaerobic respiration in animal cells and some bacteria is:

The equation for anaerobic respiration in plant cells and some microorganisms (including yeast, which is used in brewing and making bread) is:

Aerobic respiration releases more energy per molecule than anaerobic respiration – a maximum of 18 times as much. In humans, anaerobic respiration can only occur for a short period.

Animal Cells

Animal cells have the following features:

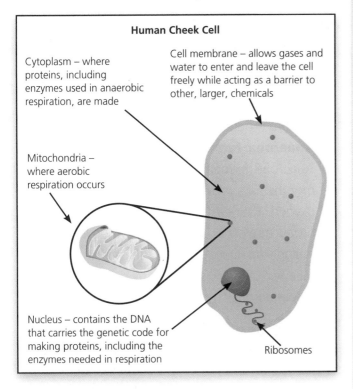

Human Cheek Cell

Cytoplasm – where proteins, including enzymes used in anaerobic respiration, are made

Cell membrane – allows gases and water to enter and leave the cell freely while acting as a barrier to other, larger, chemicals

Mitochondria – where aerobic respiration occurs

Nucleus – contains the DNA that carries the genetic code for making proteins, including the enzymes needed in respiration

Ribosomes

Microbial Cells

Microbial cells have similar structures, but with some important differences.

Bacteria

An important feature of bacteria is that they do not have any membrane-bound organelles. Therefore they do not have a nucleus or mitochondria:

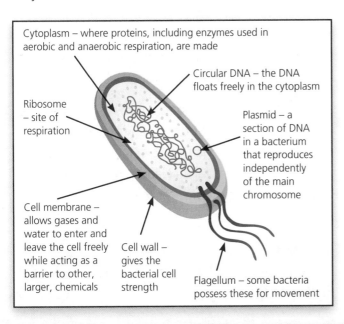

Cytoplasm – where proteins, including enzymes used in aerobic and anaerobic respiration, are made

Circular DNA – the DNA floats freely in the cytoplasm

Ribosome – site of respiration

Plasmid – a section of DNA in a bacterium that reproduces independently of the main chromosome

Cell membrane – allows gases and water to enter and leave the cell freely while acting as a barrier to other, larger, chemicals

Cell wall – gives the bacterial cell strength

Flagellum – some bacteria possess these for movement

Yeast

Yeast is a type of fungus. It is used to make bread and alcohol. Unlike bacteria, yeast does have membrane-bound organelles.

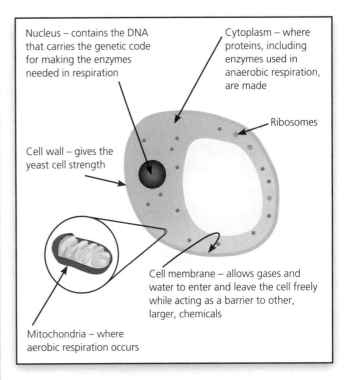

Nucleus – contains the DNA that carries the genetic code for making the enzymes needed in respiration

Cytoplasm – where proteins, including enzymes used in anaerobic respiration, are made

Ribosomes

Cell wall – gives the yeast cell strength

Cell membrane – allows gases and water to enter and leave the cell freely while acting as a barrier to other, larger, chemicals

Mitochondria – where aerobic respiration occurs

Cells Compared

The following table gives a comparison between the features of animal cells, bacteria and yeast:

	Animal Cell	Bacteria	Yeast
Cytoplasm	✓	✓	✓
Cell membrane	✓	✓	✓
Mitochondria	✓		✓
Nucleus	✓		✓
Ribosomes	✓	✓	✓
Cell wall		✓	✓
Circular DNA		✓	

Applications of Anaerobic Respiration

Biotechnology has enabled us to use the products of anaerobic respiration.

Making Bread

Yeast is added to a dough, made from flour, salt, water and other ingredients. The dough is effectively a source of glucose that is needed for anaerobic respiration.

The dough also provides an anaerobic environment for the yeast cells to grow and multiply. As they grow, they respire anaerobically – producing ethanol and carbon dioxide gas as waste products.

The gas bubbles of carbon dioxide released by the yeast are trapped in a molecule called gluten. This makes the dough expand. The ethanol quickly evaporates.

Once the dough has risen and has been baked, all that is left is the expanded structure of the bread.

Brewing Alcohol

Brewing involves a fermentation process. There are two stages:

1. **Aerobic fermentation**

   Aerobic fermentation lasts about a week and is the stage where the yeast is exposed to air and grows very rapidly on the sugar provided. Some alcohol is produced but the majority of energy is used to produce more yeast cells.

2. **Anaerobic fermentation**

   Anaerobic fermentation lasts for weeks or months. This takes place in the absence of oxygen. The yeast respires anaerobically and produces alcohol and carbon dioxide instead of multiplying. Once brewed, the carbon dioxide escapes (unless it is needed by the brewer, e.g. in producing champagne).

Approximately 70% of the fermentation takes place aerobically. The remaining 30% of the fermentation takes place anaerobically. You might wonder whether you could get more alcohol if it was left longer. This would not work because the alcohol as a waste product is poisonous to the yeast and eventually it reaches a level high enough to kill the yeast. For stronger alcoholic content, distillation has to be undertaken.

Biogas

It is now possible to introduce bacteria to biodegradable substances such as manure, sewage and household waste in landfill sites. The anaerobic digestion leads to the production of methane (an explosive gas) and carbon dioxide.

The methane can be used as a low-cost fuel. As the methane is generated from materials that were thrown away, it is regarded as being a renewable energy source. The United Nations consider the gas to be a key energy source for the future.

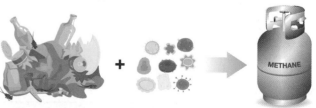

Growth and Development

Module B5 (Growth and Development)

Genetic technologies, such as stem cell research and cellular-growth control, are at the cutting edge of modern science. This module looks at:

- how organisms produce new cells
- how genes control growth and development within the cell
- how new organisms develop from a single cell.

Cells, Tissues and Organs

Cells are the building blocks of all living things. Multicellular organisms are made up of collections of cells. The cells can become **specialised** to do a particular job.

Groups of specialised cells working together are called **tissues** and groups of tissues working together are called **organs**.

For example:

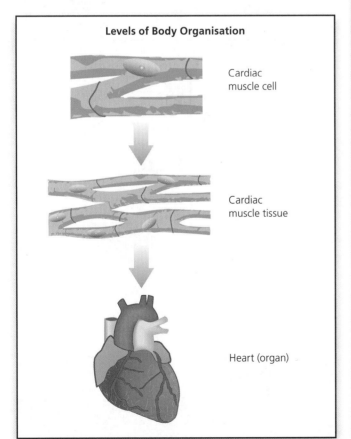

Levels of Body Organisation

Cardiac muscle cell

Cardiac muscle tissue

Heart (organ)

Mitosis

Mitosis is the process by which a cell divides to produce two new cells with identical sets of chromosomes to the parent cell. The new cells will also have all the necessary organelles.

The purpose of mitosis is to produce new cells for growth and repair and to replace old tissues.

Fertilisation

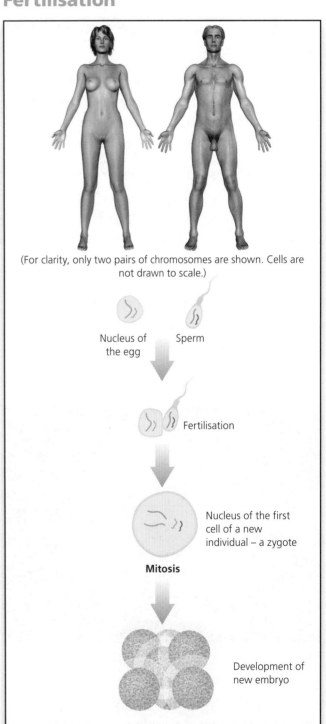

(For clarity, only two pairs of chromosomes are shown. Cells are not drawn to scale.)

Nucleus of the egg

Sperm

Fertilisation

Nucleus of the first cell of a new individual – a zygote

Mitosis

Development of new embryo

Fertilisation (Cont.)

When an egg is fertilised by a sperm it becomes a **zygote**.

(Cells are not drawn to scale – the egg is 2.72 times larger than the sperm.)

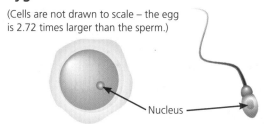

Nucleus

The zygote then divides by mitosis to form a cluster of cells called an embryo.

Up to (and including) the eight cell stage, all the cells are identical and can produce **any** sort of cell required by the organism including neurons, blood cells, liver cells, etc. They are called **embryonic stem cells**.

At the 16 cell stage (approximately four days after fertilisation), most of the cells in an embryo begin to specialise and form different types of tissue. Although the cells contain the same genetic information, the position of each cell is different relative to the others. The distribution of various proteins in the cells will also be different. So, although the genes are the same, the cells are already subtly different from one another. At the time of specialisation these differences determine what specific functions a cell will have.

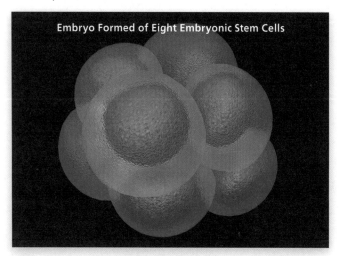

Embryo Formed of Eight Embryonic Stem Cells

Some cells remain unspecialised. These are **adult stem cells**. At a later stage they can become specialised. However, unlike embryonic stem cells, adult stem cells cannot become any type of cell.

Plant Meristems

Plants have cells that are like stem cells in animals. The cells are in areas called **meristems**. Only cells within meristems can divide repeatedly (i.e. are **mitotically active**).

Cells in the meristem are unspecialised but they can develop into any type of plant cell. Under normal hormonal conditions, this would mean that tissues such as xylem and phloem could be formed as well as organs such as leaves, roots and flowers.

There are two types of meristem: those that result in increased girth (**lateral meristems**) and those that result in increased height and longer roots (**apical meristems**):

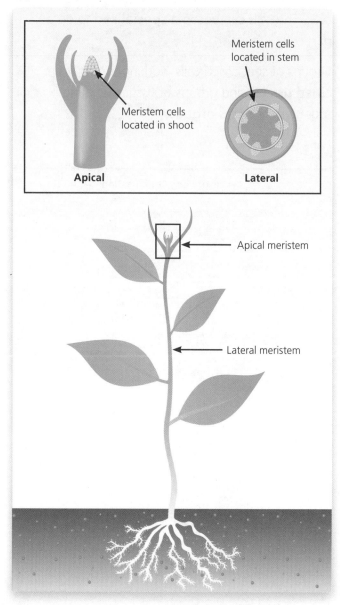

Xylem and Phloem

Xylem is made from specialised cells to transport water and soluble mineral salts from the plant roots to the stem and leaves, and to replace water lost during transpiration and photosynthesis.

Phloem is made from specialised cells to transport dissolved food made by photosynthesis throughout the plant for respiration or storage.

When a stem is deliberately cut, special plant hormones can be added. These can send messages to the meristems to start to produce roots. As the cutting already has a stem and leaves, it will then grow into a clone of the parent plant.

> **HT** There are a wide range of plant hormones. The main group that is used in horticulture is called **auxin**. Auxins mainly affect cell division at the tip of a shoot, because that is where the meristems are. Just under the tip, the cells grow in the presence of auxins, causing the stem or root to grow longer.

The growth and development of plants are affected by the environment. One example is **phototropism**. Another example is geotropism. This is where roots and shoots grow towards and away from the source of gravity. Other tropisms also exist.

Phototropism

A tropism is the term given to a response by a plant to a stimulus. So, phototropism is a response by the plant to light.

A plant's survival depends on its ability to photosynthesise. Plants therefore need strategies to detect light and to respond to changes in intensity. This is demonstrated by the way in which plants will grow towards a light source.

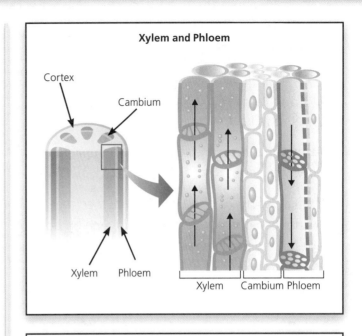

Xylem and Phloem

Cortex

Cambium

Xylem   Phloem

Xylem   Cambium Phloem

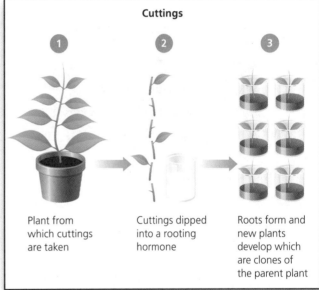

Cuttings

1 2 3

Plant from which cuttings are taken

Cuttings dipped into a rooting hormone

Roots form and new plants develop which are clones of the parent plant

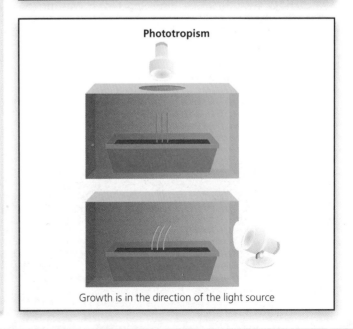

Phototropism

Growth is in the direction of the light source

How Phototropism Works

The cells furthest away from a light source grow more, due to the presence of auxin, which is sensitive to light.

Auxin is produced at the shoot tip and migrates down the shoot.

If a light source is directly overhead, then the distribution of auxin would be the same on both sides of the plant shoot.

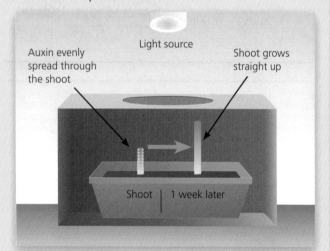

If a light source shines onto the shoot at an angle, the auxin facing the light moves to the side furthest away. The cells on the side furthest from the light source have proportionately more auxin than the closest.

As a result, the concentration of auxin on the side furthest away from the light increases, causing the cells there to elongate, and the shoot begins to bend towards the light.

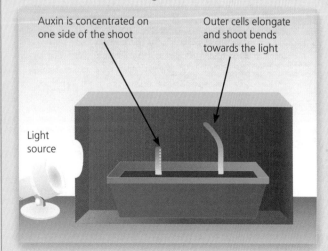

Charles Darwin carried out some simple experiments that demonstrated the role played by plant hormones produced in the shoot tip.

As we have seen, light causes the shoot to bend towards the source. If the tip of the shoot is removed or covered in opaque material then the plant will continue to grow upwards – as if the light source was not there.

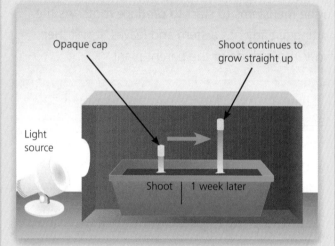

If the tip is covered with a transparent cap then it will still grow towards the light source. The same thing will happen if an opaque cylinder is wrapped around the stem leaving the tip exposed.

This experiment proves it is a substance produced in the tip that causes the cells further down the shoot to grow.

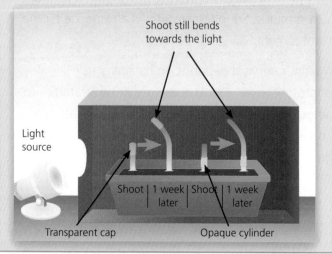

Mitosis and Growth

Mitosis leads to the production of two new cells, which are identical to each other and to the parent cell. Mitosis can only take place when a cell is ready to divide. This means that cells go through a **cell cycle**.

The cell cycle consists of a growth stage (G1) where the cell gets bigger and the number of organelles increase, then a synthesis stage (S) where the DNA is copied, followed by another very short growth stage (G2) immediately before mitosis (M phase). The cycle is a bit like a clock:

The Cell Cycle

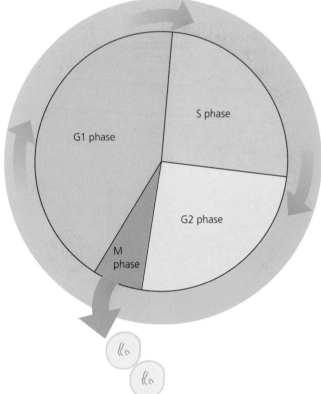

Both new cells need to have a full complement of organelles and DNA to function properly. Therefore, the number of organelles needs to increase and the DNA has to be copied.

The chromosomes are copied when the two strands of each DNA molecule separate and new strands form alongside them.

DNA Replication

The process of mitosis is shown in the following diagram:

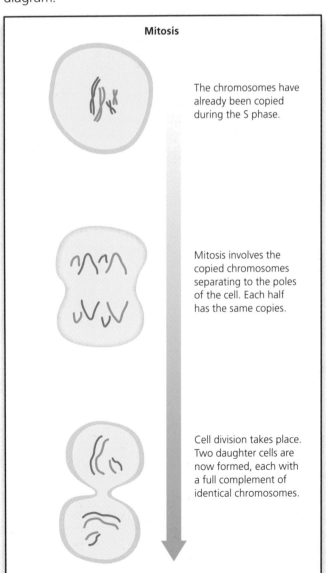

Mitosis

The chromosomes have already been copied during the S phase.

Mitosis involves the copied chromosomes separating to the poles of the cell. Each half has the same copies.

Cell division takes place. Two daughter cells are now formed, each with a full complement of identical chromosomes.

Meiosis

Meiosis only takes place in the testes and ovaries. It is a special type of cell division that produces gametes (sex cells, i.e. eggs and sperm) for sexual reproduction.

Gametes contain **half** the number of chromosomes of the parent cell. This is important because it means that when the male and female gametes fuse, the number of chromosomes will increase back to the full number. The resulting zygote has a set of chromosomes from each parent.

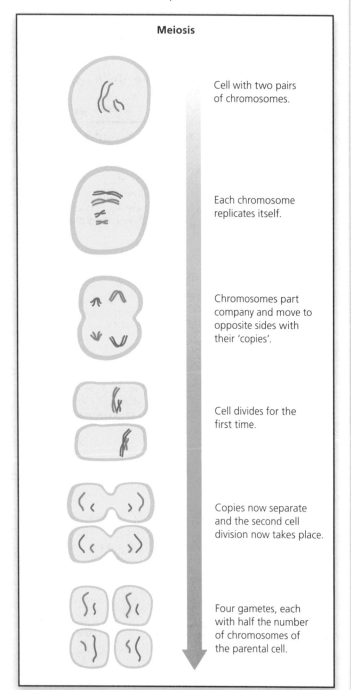

Meiosis

Cell with two pairs of chromosomes.

Each chromosome replicates itself.

Chromosomes part company and move to opposite sides with their 'copies'.

Cell divides for the first time.

Copies now separate and the second cell division now takes place.

Four gametes, each with half the number of chromosomes of the parental cell.

DNA

DNA is a nucleic acid, found in the nucleus, and is in the form of a double helix. The structure of DNA was worked out by Watson and Crick in 1952, earning them a Nobel prize for their discovery.

DNA Structure

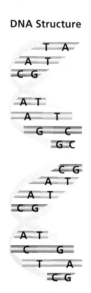

DNA

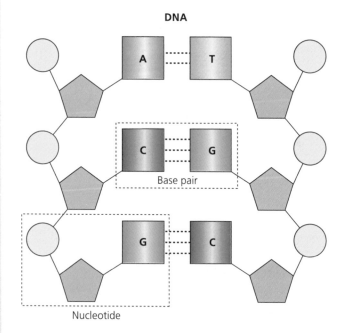

Base pair

Nucleotide

The DNA molecule has a series of bases connected together, like rungs on a ladder. The bases are always in pairs:

- Adenine (A) pairs with thymine (T).
- Guanine (G) pairs with cytosine (C).

DNA Base Pairs

The DNA is always stored within the nucleus; it does not leave. The DNA has sequences of genes, which code for proteins. However, the proteins themselves are manufactured in the cytoplasm of the cell. Therefore, there is a mechanism for transferring the information stored in the genes into the cytoplasm.

Imagine the nucleus is a reference library, from which you are not allowed to remove the books. You would need to copy down any information you needed so that you could take it away with you.

The DNA molecule is too large to leave the cell. So, the relevant section of DNA is unzipped and the instructions copied onto smaller molecules, which can pass through the nuclear membrane of the nucleus into the cytoplasm.

HT The smaller molecules are called **messenger RNA** (mRNA). These leave the nucleus and carry the instructions to the **ribosomes**, which follow the instructions to make the specific protein.

It is the **order** of bases that determines what protein is made.

HT The bases in a gene are read in threes or triplets, called a codon. Each triplet means a specific amino acid is attached. As the amino acids join together a new protein is formed.

The mRNA is the opposite of the DNA code, with the exception that mRNA has the base uracil (U) instead of T – so, if the DNA base was A, the mRNA base would be U.

The triplet codes determine the amino acid produced. The table below helps to explain how. You do not need to recall it.

DNA Bases and their Complementary mRNA Bases

DNA Base	mRNA Base
A	U
T	A
G	C
C	G

Second Base of Codon

	U	C	A	G	
U	UUU UUC — Phenylalanine, Phe, (F) / UUA UUG — Leucine, Leu, (L)	UCU UCC UCA UCG — Serine, Ser, (S)	UAU UAC — Tyrosine, Tyr, (Y) / UAA — Stop codon / UAG — Stop codon	UGU UGC — Cysteine, Cys, (C) / UGA — Stop codon / UGG — Tryptophan, Trp, (W)	U C A G
C	CUU CUC CUA CUG — Leucine, Leu, (L)	CCU CCC CCA CCG — Proline, Pro, (P)	CAU CAC — Histidine, His, (H) / CAA CAG — Glutamine, Gln, (Q)	CGU CGC CGA CGG — Arginine, Arg, (R)	U C A G
A	AUU AUC AUA — Isoleucine, Ile, (I) / AUG — Methionine, Met, (M) Start codon	ACU ACC ACA ACG — Threonine, Thr, (T)	AAU AAC — Asparagine, Asn, (N) / AAA AAG — Lysine, Lys, (K)	AGU AGC — Serine, Ser, (S) / AGA AGG — Arginine, Arg, (R)	U C A G
G	GUU GUC GUA GUG — Valine, Val, (V)	GCU GCC GCA GCG — Alanine, Ala, (A)	GAU GAC — Aspartic acid, Asp, (D) / GAA GAG — Glutamic acid, Glu, (E)	GGU GGC GGA GGG — Glycine, Gly, (G)	U C A G

First Base of Codon (left axis) · *Third Base of Codon* (right axis)

- Name of amino acid
- Abbreviated name of amino acid
- Single-letter code for amino acid

Start – when appearing for the first time, this causes the gene to be read.

Stop – when this codon is read it causes the transcription process to stop completely.

Gene Switching

It is a fundamental premise in biology that a gene codes for one specific protein. All body cells, including stem cells, contain **exactly the same genes**.

All cells have to produce proteins for growth and respiration. This means that certain genes (e.g. the gene for making the cell membrane and the gene for making ribosomes) will be switched on in all cells. The genes that are not needed will be switched off.

This can be visualised as:

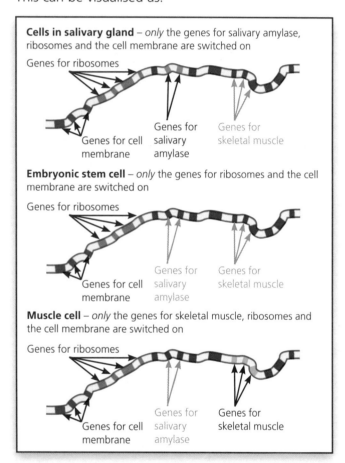

Cells in salivary gland – *only* the genes for salivary amylase, ribosomes and the cell membrane are switched on

Genes for ribosomes

Genes for cell membrane

Genes for salivary amylase

Genes for skeletal muscle

Embryonic stem cell – *only* the genes for ribosomes and the cell membrane are switched on

Genes for ribosomes

Genes for cell membrane

Genes for salivary amylase

Genes for skeletal muscle

Muscle cell – *only* the genes for skeletal muscle, ribosomes and the cell membrane are switched on

Genes for ribosomes

Genes for cell membrane

Genes for salivary amylase

Genes for skeletal muscle

When a stem cell becomes specialised, the genes for proteins specific to the new cell type will be switched on. *Any* gene in an embryonic stem cell *could* potentially be switched on, to make the specialised cell type needed.

In human DNA, there are approximately 20 000–25 000 genes, meaning a similar number of different proteins could be made. Each cell type will have a small fraction of these genes switched on.

Stem cells have the potential to produce cells needed to replace damaged tissues. For example, stem cells can be used to replace brain tissue in a patient with Parkinson's disease, or to grow new skin tissue following a burn.

Ethical Decisions

To produce the large number of stem cells needed for treatments, it is necessary to **clone** cells from five-day-old embryos. The stem cells are collected when the embryo is made up of approximately 150 cells. The rest of the embryo is destroyed. At the moment, unused embryos from IVF treatments are used for stem cell research.

There is an ethical issue as to whether it is right to use embryos to extract stem cells in this way. The debate revolves around whether or not the embryos should be classed as people.

One view is that if an embryo was left over from IVF (and would therefore never grow into a human being), it would be acceptable for stem cell research to be carried out on it as long as the parents gave their consent. However, another view is that destroying an embryo amounts to destroying a life.

The Government regulates and makes laws on such matters.

HT Mammalian Cloning

Whole animals have been cloned – Dolly the Sheep was the first to be cloned from adult skin cells. It is, however, illegal to clone a human being in this way.

Scientists can now take a mature, specialised cell and reactivate (switch on) inactive genes, effectively making it a new specialised cell type. This gives the potential to grow new tissue that is genetically the same as the patient. The benefit is that the tissue will not be rejected by the immune system of the patient. It also means that the patient does not have to take an expensive cocktail of drugs in order to prevent rejection. Such drugs can affect a patient's immune system and can stop it from fighting disease.

Neuroscience (the study of the nervous system, including the brain) is an area at the forefront of medical research, as developments and discoveries have a huge potential impact for society. This module looks at:

- how organisms respond to changes in their environment
- what reflex actions are
- how information is passed through the nervous system
- how organisms develop more complex behaviour
- how we know about the way in which the brain coordinates the senses
- how drugs affect our nervous system.

Reflexes

Living organisms can detect and respond to a **stimulus**, i.e. a change in the environment, such as light, temperature, etc.

Receptors are stimulated by the stimulus and produce a rapid, involuntary (automatic) response. In other words, the organism responds without thinking. This is called a **simple reflex**.

The simplest animals rely on reflex actions for much of their behaviour. All their movement and reactions are simple reflex responses. The reflex actions ensure that the animal will respond in a way that is most likely to result in its survival.

For example, a simple reflex response to chemicals can lead to an organism finding food quickly. A change in light level could indicate the presence of a predator, so the organism moves away.

Simple Reflexes in Humans

Newborn babies cannot think for themselves – they exhibit a range of simple reflexes for a short time after birth, which ensures that they can survive. This is a little like having a start-up disk for a computer.

The absence of the reflexes, or their failure to disappear over time, may indicate that the nervous system of the baby is not developing properly.

Some simple reflexes exhibited by newborn babies include the following:

- **Stepping reflex** – when held under its arms in an upright position with its feet on a firm surface, a baby makes walking movements with its legs.
- **Grasping reflex** – a baby tightly grasps a finger that is put in its hand.
- **Sucking reflex** – a baby sucks on a finger (or the mother's nipple) when it is put into its mouth.
- **Startle reflex** – a baby shoots out its arms and legs when startled, e.g. by a sudden loud noise.
- **Rooting reflex** – a baby turns its head and opens its mouth, ready to feed, when its cheek is stroked.

Adults also exhibit a range of simple reflexes. They are the most efficient way of quickly responding to potentially dangerous events:

- **Pupil reflex** – bright light causes muscles in the iris in the eye to contract so that the retina is not damaged.

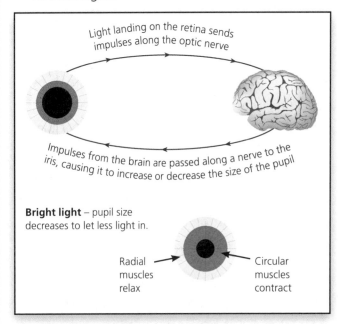

Light landing on the retina sends impulses along the optic nerve

Impulses from the brain are passed along a nerve to the iris, causing it to increase or decrease the size of the pupil

Bright light – pupil size decreases to let less light in.

Radial muscles relax

Circular muscles contract

- **Knee-jerk reflex** – when the knee is struck just below the knee cap, the leg will kick out.
- **Dropping hot object reflex** – when picking up a very hot object, the response is to throw it away in order to prevent heat damage to the hand.

Sending Signals

There are two ways of sending signals in the body. The first is via **electrical impulses** through long, wire-like cells called **neurons** (nerve cells). This method is very quick and short-lived.

Sensory neurons carry nervous impulses (electrical signals) from receptors to the central nervous system:

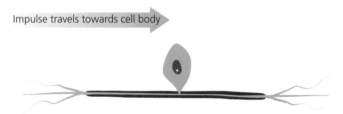

Impulse travels towards cell body

Motor neurons carry impulses from the central nervous system to effectors:

Impulse travels away from cell body

The other way signals are sent in the body is via chemicals called **hormones**, for example insulin (which controls blood sugar levels) and oestrogen (the female sex hormone), which are secreted into the blood. Chemical signals are slower than electrical impulses and they move to target organs, but their effect lasts a long time.

The nervous and hormonal communication systems are seen in larger, more complex organisms. This is a result of the **evolution** of multicellular organisms.

Detecting Changes

Nervous coordination in an animal requires the presence of one or more different **receptors** to detect stimuli. For example:

- **Light** – detected by receptors in the eyes
- **Sound** – detected by receptors in the ears
- **Changes of position** – detected by receptors for balance in the inner ear
- **Taste** – detected by receptors on the tongue
- **Smell** – detected by receptors in the nose
- **Pressure** – detected by receptors for pressure in the skin
- **Temperature** – detected by receptors for temperature in the skin

Coordinating the Response

The receptors are connected to a processing centre by sensory neurons. With simple reflexes, the processing centre is the spinal cord; the brain is not involved. The processing centre coordinates a response by sending back a message electrically via motor neurons to the **effector**, which carries out the response. This is called a **spinal reflex arc**.

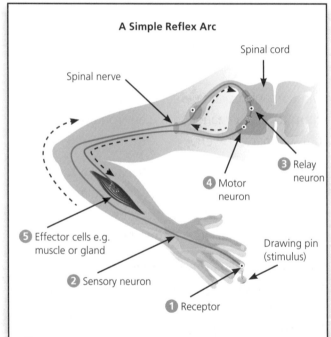

A Simple Reflex Arc

Spinal cord

Spinal nerve

③ Relay neuron

④ Motor neuron

⑤ Effector cells e.g. muscle or gland

Drawing pin (stimulus)

② Sensory neuron

① Receptor

① A receptor is stimulated by the drawing pin (stimulus)…

② …causing impulses to pass along a sensory neuron into the spinal cord.

③ The sensory neuron synapses with a relay neuron, by-passing the brain.

④ The relay neuron synapses with a motor neuron, sending impulses down it…

⑤ …to the muscles (effectors) causing them to contract in response to the sharp drawing pin.

The arrangement of neurons into a fixed pathway in a spinal reflex arc means that the responses are automatic and, hence, very rapid because no processing is required.

If the signal had to travel to the brain and be processed before action was taken, then, by the time the response arrived, it may be too late.

Receptors and Effectors

Receptors and effectors can form part of complex organs.

Muscle Cells in Muscle Tissue

The specialised cells that make up muscle tissues are effectors. Impulses travel along motor neurons and terminate at the muscle cells. These impulses cause the muscle cells to contract.

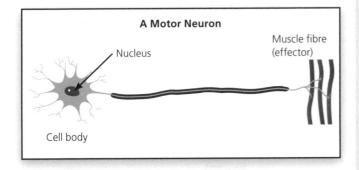

A Motor Neuron

Nucleus

Muscle fibre (effector)

Cell body

Light Receptors in the Retina of the Eye

The eye is a complex sense organ. The lens focuses light onto receptor cells in the retina, which are sensitive to light. The receptor cells are then stimulated and send electrical impulses along the sensory neurons to the brain.

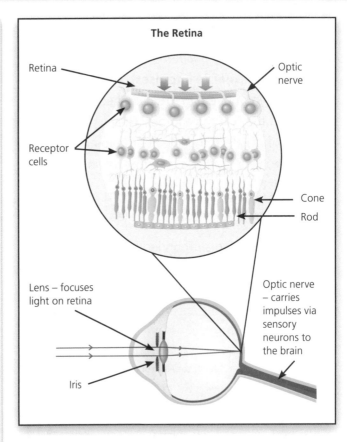

The Retina

Retina

Optic nerve

Receptor cells

Cone

Rod

Lens – focuses light on retina

Optic nerve – carries impulses via sensory neurons to the brain

Iris

Hormone Secreting Cells in a Gland

The hormone secreting cells in glands are effectors. They are activated by an impulse, which travels along a motor neuron from the **central nervous system** and terminates at the gland. The impulse triggers the release of the hormone into the bloodstream, which transports it to the sites where it is required.

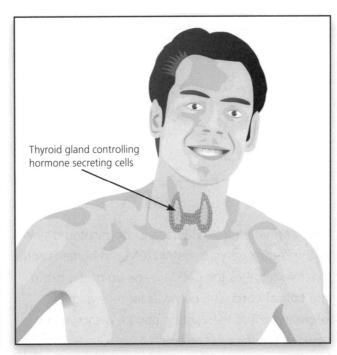

Thyroid gland controlling hormone secreting cells

The Structure of Neurons

Neurons are specially adapted cells that can carry electrical signals (**nerve impulses**). They are elongated (lengthened) to make connections from one part of the body to another. They have branched endings, which allow a single neuron to act on many other neurons or effectors, e.g. muscle fibres.

In motor neurons, the cytoplasm forms a long fibre surrounded by a cell membrane called an **axon**.

Some axons are also surrounded by a fatty sheath, which insulates the neuron from neighbouring cells (a bit like the plastic coating on a copper electrical wire) and increases the speed at which the nerve impulse is transmitted.

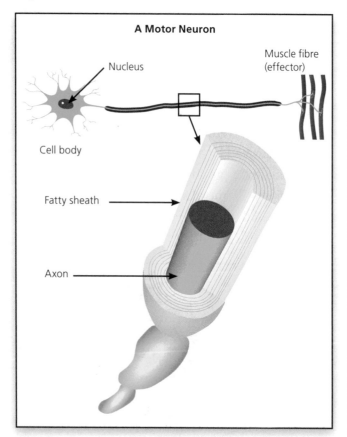

A Motor Neuron

Nucleus

Muscle fibre (effector)

Cell body

Fatty sheath

Axon

The Central Nervous System

The information from neurons is coordinated overall by the central nervous system (CNS). In humans and other vertebrates the CNS is made up of the **brain** and **spinal cord**. The pathway for receiving information and then acting upon it is shown in the flowchart in the next column.

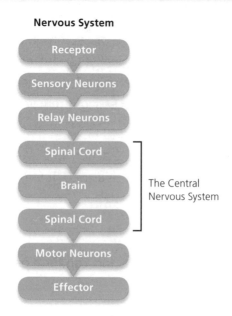

Nervous System

Receptor

Sensory Neurons

Relay Neurons

Spinal Cord

Brain — The Central Nervous System

Spinal Cord

Motor Neurons

Effector

The CNS is connected to the body via sensory and motor neurons, which make up the **peripheral nervous system (PNS)**. The PNS is the second major division of the nervous system. Its sensory and motor neurons transmit messages all over the body, e.g. to the limbs and organs. They also transmit messages to and from the CNS.

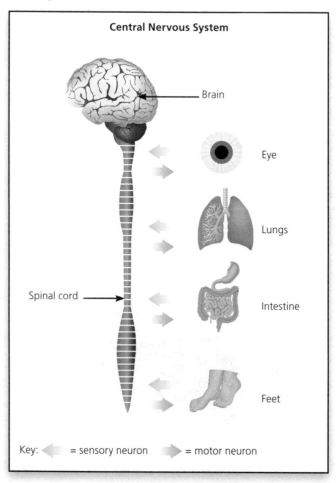

Central Nervous System

Brain

Eye

Lungs

Spinal cord

Intestine

Feet

Key: ◄ = sensory neuron   ► = motor neuron

Synapses

Synapses are the gaps between adjacent neurons. They allow the brain to form interconnected neural circuits. The human brain contains a huge number of synapses. There are approximately 1000 trillion in a young child. This number decreases with age, stabilising by adulthood. The estimated number of synapses for an adult human varies between 100 and 500 trillion.

HT When an impulse reaches the end of a sensory neuron, it triggers the release of chemicals, called **transmitter substances**, into the synapse. They diffuse across the synapse and then bind with specific receptor molecules on the membrane of a relay neuron.

The receptor molecules will only bind with specific chemicals to initiate a nerve impulse in the relay neuron, so the signal can continue on its way. Meanwhile, the transmitter substance is reabsorbed back into the sensory neuron, to be used again.

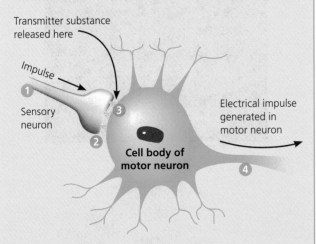

The sequence is as follows:
1. Electrical signal (nerve impulse) moves through sensory neuron.
2. Transmitter substances are released into the synapse.
3. Transmitter substances bind with receptors on the motor neuron.
4. Electrical signal (nerve impulse) is now sent through motor neuron.

Drugs and the Nervous System

Many drugs, such as Ecstasy, beta blockers and Prozac, cause changes in the speed at which nerve impulses travel to the brain, speeding them up or slowing them down. Sometimes false signals are sent.

Drugs and toxins can prevent impulses from travelling across synapses or they can cause the nervous system to become overloaded with too many impulses. For example, Ecstasy (an illegal recreational drug) and beta blockers (a drug used to prevent heart attacks) both affect the transmission of nerve impulses.

HT The drug Ecstasy, scientifically known as MDMA, in the nervous system affects a transmitter substance called **serotonin**. Serotonin can have mood-enhancing effects, i.e. it is associated with feeling happy.

Serotonin passes across the brain's synapses, landing on receptor molecules. Serotonin that is not on a receptor is absorbed back into the transmitting neuron by the transporter molecules. Ecstasy blocks the sites in the brain's synapses where the chemical serotonin is removed.

As a result, serotonin concentrations in the brain increase and the user experiences feelings of elation. However, the neurons are harmed in the process and a long-term consequence of taking Ecstasy can be memory loss.

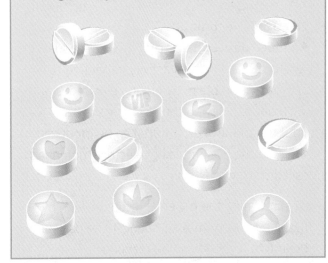

The Cerebral Cortex

The **cerebral cortex** is the part of the brain most concerned with intelligence, memory, language and consciousness (our being aware of our own thinking and existence). Scientists have used a variety of methods to map the different regions of the cerebral cortex:

- **Physiological Techniques**

  Damage to different parts of the brain can produce different problems, e.g. long- and short-term memory loss, paralysis in one or more parts of the body, speech loss, etc. Studying the effects of accidents or illnesses, as well as by directly stimulating the brain with electrical impulses, has led to an understanding of which parts of the brain control different functions:

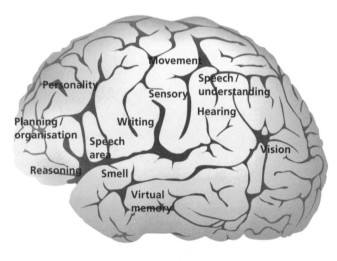

N.B. You do not need to learn the different parts of the brain.

- **Electronic Techniques**

  An **electroencephalogram** (EEG) is a visual record of the electrical activity generated by neurons in the brain. By placing electrodes on the scalp and amplifying the electrical signals picked up through the skull, a trace can be produced showing the rise and fall of electrical potentials called brain waves.

  By stimulating the patient's receptors (e.g. by flashing lights or making sounds), the parts of the brain that respond can be mapped.

Magnetic Resonance Imaging (MRI) scanning is a technique that produces images of cross-sections of the brain, showing its structure. The computer-generated picture uses colour to represent different levels of electrical activity. The activity in the brain changes depending on what the person is doing or thinking.

In 2010, patients in a persistent vegetative state after being involved in serious accidents were tested using MRI. It was shown that, although they were outwardly comatose, they could answer questions by imagining playing tennis to represent 'yes' and sitting down to represent 'no'. The parts of the brain that show increased electrical activity when playing or imagining playing tennis are different to those that show electrical activity when doing or thinking about something else.

Composite Image of MRI Scans of a Patient Showing Different Sections of the Head and Brain

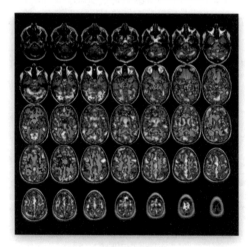

MRI Scanner

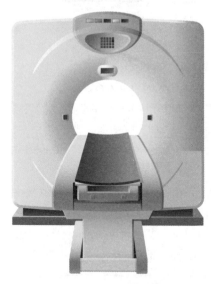

Conditioned Reflexes

Although they are not conscious actions, reflex responses to a new stimulus can be learnt. Through a process of conditioning, the body learns to produce a specific response when a certain stimulus is detected. Conditioning works by building an association between the new stimulus (the **secondary stimulus**) and the stimulus that naturally triggers the response (the **primary stimulus**). This resulting reflex is called a **conditioned reflex action**.

The effect was discovered at the beginning of the 20th century by a Russian scientist named Pavlov, who received a Nobel prize for his work.

Pavlov observed that whenever a dog sees and smells a piece of meat, it starts to salivate (produce saliva). In his experiment, a bell was rung repeatedly whenever meat was shown and given to the dog. Eventually, simply ringing the bell, without any meat present, caused the dog to salivate.

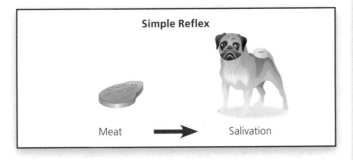

Simple Reflex

Meat ➡ Salivation

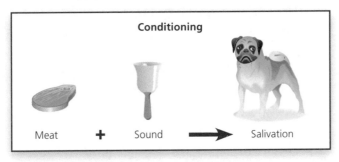

Conditioning

Meat **+** Sound ➡ Salivation

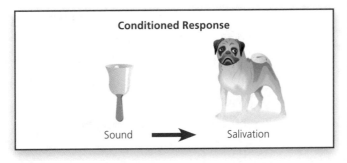

Conditioned Response

Sound ➡ Salivation

Another example of a conditioned reflex is when, after being stung by a wasp, you associate the yellow and black stripes with the painful sting. The next time you see a wasp (or similar insect such as a hoverfly), you feel fear.

HT In a conditioned reflex, the final response has no direct connection to the stimulus.

For example, the ringing bell in Pavlov's experiment is a stimulus that has *nothing* to do with feeding, it is just a sound. However, the association between the sound of the bell and the meat was strong enough to induce the dog to salivate.

Some conditioned reflexes can increase a species' chance of survival:

Example
The caterpillar of the cinnabar moth is black and orange in colour, to warn predators that it is poisonous. After eating a few cinnabar caterpillars, a bird will start to associate the colours with a very unpleasant taste and will avoid eating anything that is black and orange in colour (even if they are not actually poisonous).

In some situations, the brain can override or modify a reflex action.

For example, when the hand comes into contact with something hot, the body's natural reflex response is to pull away or drop the object.

However, you might know an object is hot but still want to pick it up, e.g. a hot potato or a dinner plate. To allow this, the brain sends a signal via a neuron to the motor neuron in the reflex arc, modifying the reflex so that you do not drop the potato or plate.

Development of the Brain

Mammals have a complex brain that contains billions of neurons. This enables them to learn from experience, including how to respond to different situations, e.g. social behaviour (dogs can be trained to be passive or aggressive).

The evolution of one particular mammal, *Homo sapiens*, led to the development of a larger brain than in other animals. Early humans could use tools, coordinate hunting and formulate plans about what might happen in the future. Having a larger brain meant that early humans were more likely to survive and reproduce, passing on the genes for producing a larger brain.

In mammals, neuron pathways are formed in the brain during development. The way in which the animal interacts with its environment determines what pathways are formed.

HT It is the huge variety of pathways available that makes it possible for the animal to adapt to new situations.

During the first few years after birth, the brain grows very rapidly. As each neuron matures it sends out multiple branches, increasing the number of synapses.

At birth, in humans, the cerebral cortex has approximately 2500 synapses per neuron. By the time an infant is two or three years old, the number of synapses is approximately 15 000 synapses per neuron.

Each time an individual has a new experience, a different pathway between neurons is stimulated. Every time the experience is repeated, the pathway is strengthened. Pathways that are not used regularly are eventually deleted. Only the pathways that are activated most often are preserved.

These modifications mean that certain pathways of the brain become more likely to transmit impulses than others and the individual will become better at a given task. This is why certain tasks may be learned through repetition, e.g. riding a bike, revising for an exam, learning to ski or learning to play a musical instrument.

HT Feral Children

If neural pathways are not used then they are deleted. There is evidence to suggest that because of this, if a new skill (e.g. learning a language) has not been learned by a particular stage in development, an animal or child might not be able to learn it in the same way as normal.

One example of evidence showing this comes from the study of so-called feral children ('feral' means 'wild').

Feral children are children who have been isolated from society in some way, so they do not go through the normal development process. This can be deliberate (e.g. inhumanely keeping a child in a cellar or locked room), or it can be accidental (e.g. through being shipwrecked).

In the absence of any other humans, the children do not ever gain the ability to talk (or they lose any ability they had already gained) other than making rudimentary grunting noises. Learning a language later in life is a much harder and slower process.

Child Development

After children are born, there are a series of milestones that can be checked to see if development is following normal patterns.

If these milestones are missing or are late, it could mean that there are neurological problems or that the child is lacking stimulation (is not being exposed to the necessary experiences).

For example, at three months, babies should be able to lift their heads when held to someone's shoulder or grasp a rattle when it is given to them. By about 12 months, babies should be able to hold a cup and drink from it, and walk when one of their hands is held.

Memory

Memory is the ability to store and retrieve information. Scientists have produced models to try to explain how the brain facilitates this but, so far, none have been able to provide an adequate explanation. Current models are limited.

Verbal memory (words and labels) can be divided into **short-term** and **long-term memory**. Short-term memory is capable of storing a limited amount of information for a limited amount of time (roughly 15–30 seconds). Long-term memory can store a seemingly unlimited amount of information indefinitely.

When using short-term memory, it is currently thought that up to seven (+/- two) separate pieces of information can be stored (e.g. for a number with eight digits like 24042003, each number could fill a slot if eight were available).

This capacity can be increased by **chunking** the information, i.e. putting it into smaller chunks. For example, the number 201054738087 could be stored as (2010), (5473) and (8087), using only three of the seven units of storage.

Long-term memory is where information is stored in the brain through repetition, which strengthens and builds up neuron pathways.

Remember, Remember...

Humans are more likely to remember information when:
* it is repeated (especially over an extended period of time), e.g. going over key points several times as a method of revising for exams
* there is a strong stimulus associated with it, such as colour, light, smell or sound (the more senses that are involved, the better)
* there is a pattern to it (or if a pattern can be artificially imposed upon it). For example, people may find remembering the order of how organisms are classified difficult, so a sentence where there is a logical relationship between the

words (unlike the names of the taxonomical groups themselves) can be remembered as a prompt instead: **K**ids **P**refer **C**heese **O**ver **F**ried **G**reen **S**pinach. The first letter of each word can trigger the biological term, as in: Kingdom, Phylum, Class, Order, Family, Genus, Species. This is called a **mnemonic** and is an example of an artificially imposed pattern. Another type of mnemonic is one that helps you to remember a rule. For example, 'tea with two sugars' is a mnemonic for the correct number of t's and s's when spelling 'potassium'.

Memory Models

There are a variety of models used to explain how memory works. The **magical number seven rule, +/- two model**, described opposite, was developed in 1956. In the late 1960s, the **multi-store model** was proposed by Atkinson and Shiffrin. This linked short- and long-term memory with a very quick sensory memory, lasting only 1–2 seconds. This model is useful as it suggests a relationship between memory centres for storage and retrieval, together with an explanation as to how repetition helps people to remember things. It also explains why people forget.

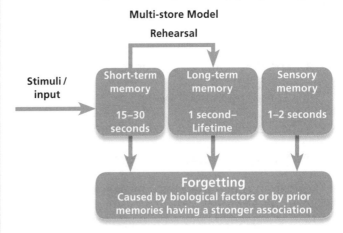

Multi-store Model

Biologists are still creating models to explain memory. Models are used to try to explain experimental results. Those that best explain and fit the data are maintained; those that do not are discarded.

However, models will always be limited as they are not the real brain and memory, only a representation of how we think it works.

Exam Practice Questions

1. On the axes below draw a graph that has the optimum temperature for enzyme activity as being 35°C. **[3]**

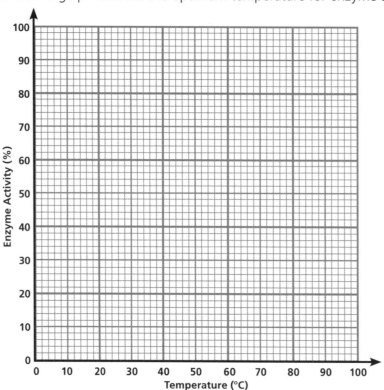

2. Explain the differences between aerobic and anaerobic respiration. **[6]**

   ✏️ *The quality of written communication will be assessed in your answer to this question.*

3. Which of the following are examples of osmosis?
   Put ticks (✔) in the boxes next to the **three** correct answers. **[3]**

   Water evaporating from leaves ☐

   Water moving from plant cell to plant cell ☐

   Mixing pure water and sugar solution ☐

   Ink spreading through water ☐

   A pear losing water in a concentrated salt solution ☐

   Water moving from blood to body cells ☐

   Sugar absorbed from the intestines into the blood ☐

4. What is needed before biologists will accept a new theory on memory?
   Put ticks (✔) in the boxes next to the **three** correct answers. **[2]**

   The theory must…

   …have a scientific mechanism. ☐   …be easy to understand. ☐

   …be from a qualified biologist. ☐   …be published. ☐

   …have data with high variability. ☐   …be repeatable. ☐

5 At what stage do cells in an embryo start to become specialised? [1]

6 This question is about responding to stimuli.
(a) Which way will a seedling in the plant pot grow? [1]

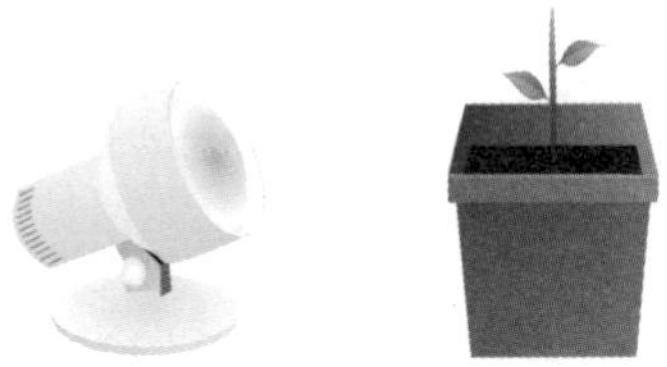

(b) Fill in the words to complete the passage below, which describes what happens when organisms respond to stimuli. [2]

Animals respond to stimuli in order to keep themselves in conditions that will ensure

their .... These responses are coordinated by the ....

7 This question is about reflexes.
(a) Name **three** reflexes shown by newborn babies. [3]

(b) Explain how simple reflexes enable organisms to survive. [6]
✎ *The quality of written communication will be assessed in your answer to this question.*

(c) What is the name given to reflexes that are learned? [1]

HT

8 Write the chemical equation for aerobic respiration. [2]

9 Write the complementary mRNA sequence for the following stretch of DNA:

AATAGCCGCCAATTAGGC [1]

10 Four students are talking about Ecstasy.

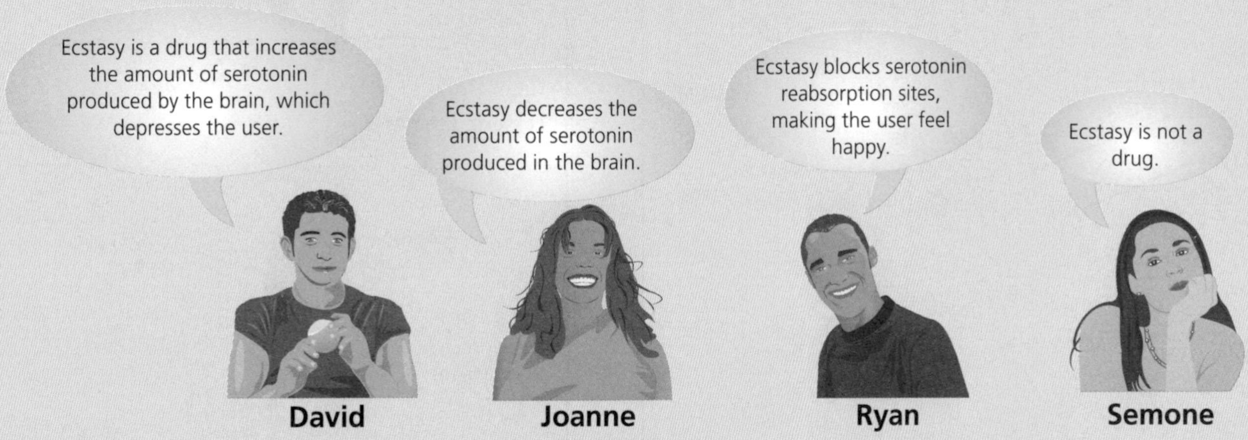

Ecstasy is a drug that increases the amount of serotonin produced by the brain, which depresses the user.

Ecstasy decreases the amount of serotonin produced in the brain.

Ecstasy blocks serotonin reabsorption sites, making the user feel happy.

Ecstasy is not a drug.

David　　　　**Joanne**　　　　**Ryan**　　　　**Semone**

Who is giving the most accurate statement? [1]

Module B7 (Peak Performance)

To compete in events such as the London 2012 Olympics, athletes have to train thoroughly to ensure their bodies are able to perform at their peak. This topic looks at:

- the structure and function of joints
- the importance of physical exercise
- how we determine a person's fitness
- the treatment of physical injuries such as sprains
- the importance and role of the circulatory system
- how temperature and water levels are maintained.

Vertebrates

Vertebrates are animals that have an internal skeleton. Those that do not have an internal skeleton are called **invertebrates**. In humans and other vertebrates, the skeleton has two functions:

- **Support** – the skeleton enables us to stand as well as enclosing important organs for protection, e.g. the brain is enclosed by the skull, and the ribs enclose and protect the heart and lungs

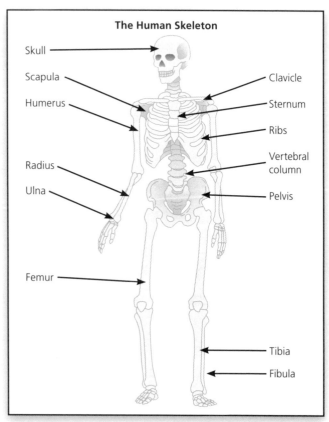

N.B. You do not need to be able to label the skeleton.

- **Movement** – the skeleton enables complex movement, from standing to sitting and from walking to running. Muscles are attached to various bones that enable arms and legs to act as levers to enable those movements. These movements could not happen in organisms without an internal skeleton.

Joint Movement

Moving your arm requires more than just a single muscle. Muscles can only move bones by **contracting**. So, two muscles working in opposition to one another are needed to move an arm up and down, i.e. one muscle contracts while the other muscle relaxes. In other words, they work in **antagonistic pairs**.

For example:

- to lift the lower arm, the biceps contracts and the triceps relaxes
- to lower the arm, the triceps contracts and the biceps relaxes.

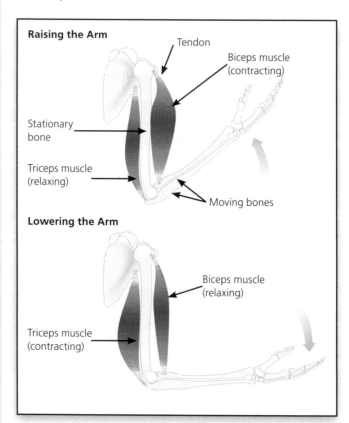

If a tendon connecting the triceps to the bone was cut, the triceps would not be able to contract and the arm would remain in the up position.

Structure and Function of Joints

A joint is where two bones meet and work together. The bones have to be connected in some way that allows them to move while also staying in the same place relative to each other. This requires a number of different proteins to work together.

In a joint, such as the knee joint shown below, there is a smooth layer of a stiff, inflexible protein called **cartilage** and a viscous fluid called **synovial fluid**. The purpose of cartilage and synovial fluid is to reduce friction and wear and tear between bones.

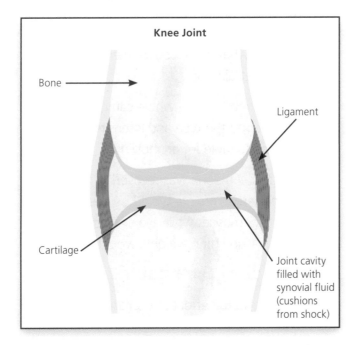

Knee Joint

Bone

Ligament

Cartilage

Joint cavity filled with synovial fluid (cushions from shock)

Ligaments are an elastic, fibrous tissue, made from collagen, which join bones together and stabilise them while still allowing movement.

Tendons are a tough fibrous tissue, made from collagen, which connect a muscle to a bone. Their purpose is to transmit the forces between the two. The force is only a pulling force.

Taken together, the specific properties of each part of a joint enable it to function correctly. If there is a problem with one part, the joint will not work correctly.

For example, if a tendon snapped then there would be no way of moving the bone. If knee cartilage was damaged then moving the leg would be a painful process, rather than a smooth one, due to the

friction caused. For footballers, a torn knee ligament is a devastating injury that stops them playing for a season, and can possibly end their career.

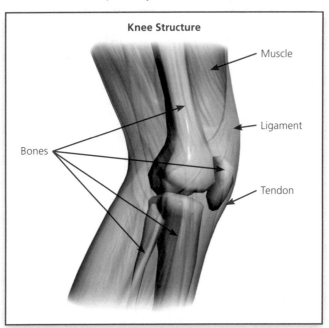

Knee Structure

Muscle

Ligament

Bones

Tendon

Medical and Lifestyle History Assessment

When someone wishes to carry out an exercise regime, e.g. to train for a sport or to treat a condition such as obesity, it is important that the medical and lifestyle history of the person is examined beforehand. This helps to ensure that the exercise regime is effective and safe, e.g. it does not make any health problems worse or does not trigger any other ones.

Practitioners who might develop an exercise regime include fitness instructors, doctors, nurses and physiotherapists, i.e. people who have had special scientific or medical training.

Developing an Exercise Regime

Important factors that a practitioner needs to consider when developing an exercise programme include the following:

- **Symptoms** – visible or noticeable effects of a disease or condition on the body are usually noticed by the patient and can be used to identify problems. For example, someone who experiences a burning pain near their elbow after they have been lifting objects may have tendonitis. With those symptoms, it would not be sensible to exercise that part of the body.
- **Current medication** – different medicines can sometimes conflict with each other and affect how the body responds under certain conditions, e.g. stress.
- **Alcohol consumption** – high levels of alcohol consumed on a regular basis can cause physical problems. Body weight can increase to unhealthy levels and the kidney and liver can be damaged.
- **Tobacco consumption** – there are many diseases and disorders directly related to smoking, such as lung cancer, bronchitis and emphysema. Smokers have a higher risk of developing heart disease and high blood pressure. If a person exercises too much with these conditions, it could trigger heart failure.
- **Level of physical activity** – in general, the more exercise a person takes, the healthier they are. An individual who does little exercise is likely to get tired quicker, may have weak bones, and may

have problems sleeping and concentrating. In such a case, an exercise regime would have to take account of the fact that the individual is not used to putting their body under physical stress.

- **Family medical history** – some medical conditions can be genetic and therefore inherited. Genetic conditions such as heart disease need to be identified. If a family member is affected then the person undertaking the exercise programme may also be.
- **Previous treatments** – if a person is undertaking an exercise regime, it is useful to know what treatments they have had in the past. This can be a guide to designing the new regime, building on what previously worked and avoiding anything that did not.

Using this information as a whole can guide a physical trainer on the type and intensity of exercises that would be suitable for an individual.

For example, if someone was recovering from back surgery the exercise regime would be designed to strengthen back muscles while avoiding stressing them. Exercises involving lifting weights would be avoided.

The Effect of Exercise

The graph shows the effect of exercise on heart rate and blood pressure.

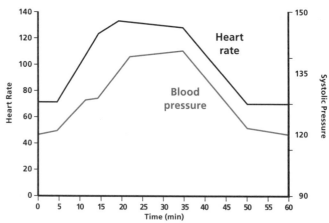

The exercise started after 5 minutes. As the exercise increased, the heart rate increased – more blood had to be pumped to the muscles and lungs. The exercise stopped at 35 minutes. It then took 15 minutes for the heart rate and blood pressure to return to normal.

Body Mass Index

A person's height and mass can be used to determine their **body mass index (BMI)**. The BMI is a guideline that helps to identify whether a person is a healthy mass. It is calculated by the following formula:

$$\text{Body mass index} = \frac{\text{Body mass (kg)}}{[\text{Height (m)}]^2}$$

Example

Arwen is 1.85m tall and weighs 70kg. What is her BMI?

$$\text{BMI} = \frac{70}{(1.85)^2} = \textbf{20.5}$$

A healthy BMI is between 18.5 and 25.

It can be seen from the body mass graph below, another way of viewing BMI, that Arwen is in the middle of the 'OK' band.

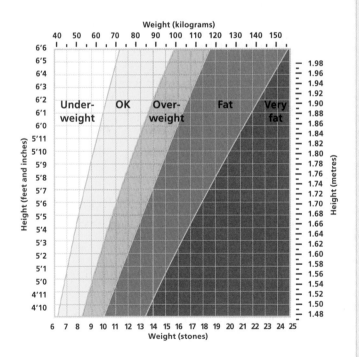

Measuring Body Fat

The BMI does not take into account the proportion of body fat. So, along with the BMI, a physical test should be carried out to determine the amount of body fat. This can be done using callipers to measure the thickness of folded skin.

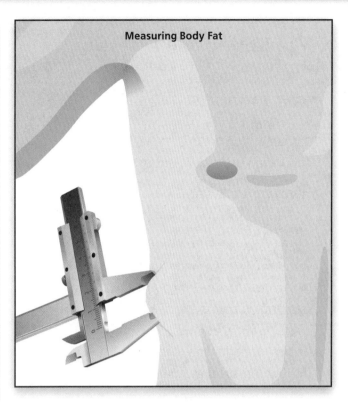

Measuring Body Fat

The table below shows approximately how much fat is present if a skin calliper reading is taken.

Skin Thickness (mm)	% Body Fat	
	Male	Female
6	5–9	8–13
13	9–13	13–18
19	13–18	18–23
25	18–22	23–28
38	22–27	28–33

Fat is essential for the body. If the level of fat drops to below 6% in a man and 14% in a woman, then natural body processes that rely on fat are affected.

The amount of body fat required depends on what the person is doing. For example, a female athlete would need 14–20% and a male athlete 6–13%.

Generally, a person is regarded as obese if their body fat is more than 25% (in males) or more than 32% (in females), although age will alter this boundary.

An alternative to callipers is to use the conductivity of the skin to indicate the proportion of body fat.

If both the BMI and the body-fat proportion are high, the person needs to lose weight.

Monitoring and Assessing Progress

A treatment or fitness-training regime needs to be monitored to check that it is having the desired effect. It can then be modified depending on the patient's progress.

This is dependent on the accuracy of the monitoring techniques and the repeatability of the data.

For example, measuring weight loss can be checked simply by standing on bathroom scales and checking the mass. However, our bodies lose and gain mass throughout the day, from eating, drinking and going to the toilet.

If a person checked their mass too often it would generate variable data that could not be relied upon. It would be better to measure body mass once every few days and take the measurement at the same time of the day. This would make the data more reliable.

Injuries From Exercise

The human body can withstand a lot of exercise. However, **excessive exercise** (over-exertion or not being properly prepared for exercise) can put the body under a lot of strain, which can lead to injuries. Injuries include sprains, dislocations and torn ligaments or tendons.

Dislocation

A sudden, severe impact can cause certain joints to become **dislocated**. A dislocation is where a joint becomes misaligned or the bones are disconnected from the joint, e.g. the leg is displaced from the socket in the hip.

The unusual position of the bones is very painful and can also result in torn ligaments and tendons. Dislocated joints can often be mistaken for broken bones because they produce similar pain, misshapen body parts and severe swelling.

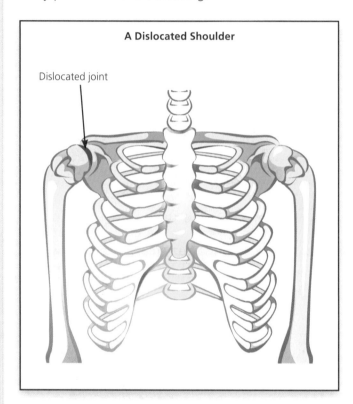

A Dislocated Shoulder

Dislocated joint

Sprains

A sprain is where an activity causes a stretch in a ligament beyond its natural capacity. Ankles, knees and wrists are all vulnerable to strains.

A sprained ankle can occur when the foot turns inward because this puts extreme tension on the ligaments of the outer ankle.

A sprained knee can be the result of a sudden twist. A wrist can be sprained by falling on an outstretched hand.

The symptoms of sprains include the following:
- Swelling – due to fluid building up at the site of the sprain.
- Pain – the joint hurts and may throb. The pain can increase if the injured area is pressed or moved in certain directions, or if weight is put on it.
- Redness and warmth – caused by the increased blood flow to the injured area.

When someone suffers a sprain the priority is to reduce swelling and pain, and aid rapid recovery and rehabilitation. The treatment follows the principle of **RICE**:

- **R**est – the patient rests and does not put pressure on the injured part of the body.
- **I**ce – should be placed on the injured part for short periods to reduce swelling and bleeding (e.g. an ice pack, wrapped in fabric to prevent ice burns).
- **C**ompression – gentle pressure should be applied with a bandage to reduce the build-up of fluid that causes swelling.
- **E**levation – the injured part should be raised to reduce blood pressure and aid blood flow into the injured area.

Torn Ligaments and Tendons

A particularly severe sprain could mean that a ligament has been torn. Tendons can also be torn.

A torn ligament or tendon is very painful. Recovery can take a long time. The blood supply to ligaments and tendons is poor compared with other parts of the body, so the materials needed for repair (proteins, etc.) are slower to arrive.

A further consequence is that the bones connected to the ligament or tendon may not be in the correct position. If the tendon is torn, there is no way the limb can move – the muscle is effectively detached. Surgery may be required.

Physiotherapy

A physiotherapist specialises in the treatment of skeletal-muscular injuries. Physiotherapists understand how the body works and can help a patient to re-train or reuse a part of the body that is not functioning properly. This is normally achieved through various exercises to strengthen muscles that may have become weakened.

There are many different exercises and it is the job of the physiotherapist to choose the best course of treatment for each patient. For example, an injured leg could be treated with the following exercise programme:

- Warming up the joint by riding a stationary exercise bicycle, then straightening and raising the leg.
- Extending the leg while sitting (a weight may be worn on the ankle for this exercise).
- Raising the leg while lying on the stomach.
- Exercising in a swimming pool (walking as fast as possible in chest-deep water, performing small flutter kicks while holding onto the side of the pool, and raising each leg to 90° in chest-deep water while pressing the back against the side of the pool).

The Circulatory System

Microscopic organisms do not need a circulatory system. Dissolved food and gases can diffuse through the cells easily. Once a certain size of organism is reached, diffusion is no longer effective. Cells that are out of reach of the gases and food cannot carry out respiration.

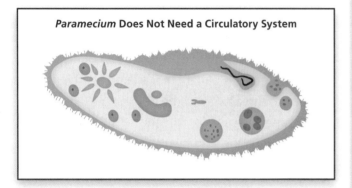

Paramecium Does Not Need a Circulatory System

For larger organisms (e.g. an earthworm), a circulation system is needed. This involves a simple pump (a heart) forcing blood around the body. It ensures that all cells can get their dissolved nutrients and gases from the blood.

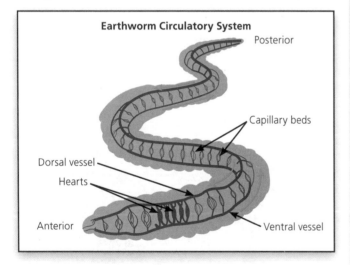

Earthworm Circulatory System

Posterior

Capillary beds

Dorsal vessel

Hearts

Anterior

Ventral vessel

For the largest animals (e.g. mammals), a more efficient heart is needed as the organisms are larger still. The heart of these animals is effectively a **double pump**. Part of the heart pumps **deoxygenated** (without oxygen) blood to the lungs to pick up oxygen and then returns the now **oxygenated** blood to the heart to be pumped around the rest of the body. This type of circulation is called **double circulation**.

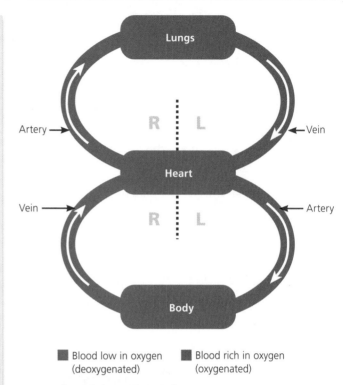

Lungs

Artery

R | L

Vein

Heart

Vein

R | L

Artery

Body

■ Blood low in oxygen (deoxygenated) ■ Blood rich in oxygen (oxygenated)

The Heart

The **heart** is a muscular organ in the circulatory system that beats automatically, pumping blood around the body. The rate at which the heart beats varies according to stress, exertion and disease.

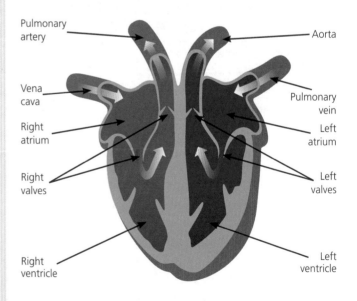

Pulmonary artery

Aorta

Vena cava

Pulmonary vein

Right atrium

Left atrium

Right valves

Left valves

Right ventricle

Left ventricle

Most of the heart is made of muscle. The left side of the heart is more muscular than the right side because it pumps blood around the whole body, whereas the right side only pumps to the lungs.

Features of the Heart

The heart has the following features:

- Two **atria**, which are the smaller, less muscular upper chambers that receive blood coming back to the heart from the veins.
- Two **ventricles**, which are the larger, more muscular lower chambers.
- The **vena cava** is a larger vein that returns blood from the body into the right atrium.
- The **pulmonary arteries** (one for each lung) transport blood to the lungs. Blood travels from the right ventricle to the lungs. The pulmonary arteries are the only arteries to carry deoxygenated blood.
- The **pulmonary vein** returns blood from the lungs to the left atrium. It is the only vein that transports oxygenated blood.
- The **aorta** is the largest artery in the body, taking oxygenated blood at high pressure to the whole body from the left ventricle.

Looking at an external view of the heart, you can see that there are arteries on the outside. The coronary arteries supply the heart with blood so that the heart cells can respire.

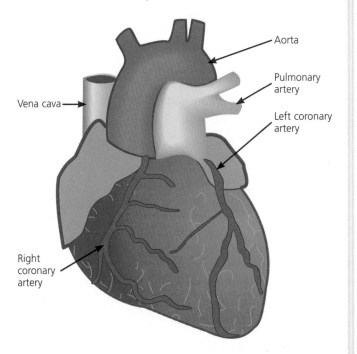

The pressures that build up inside the atria and ventricles are very high. Valves between each chamber and the arteries leaving the heart prevent the back flow of blood (which would stop the circulatory system from working).

Valves are also found in veins in the rest of the body. Blood must only travel in one direction – if it moves back, it causes the valve to close (preventing further back flow).

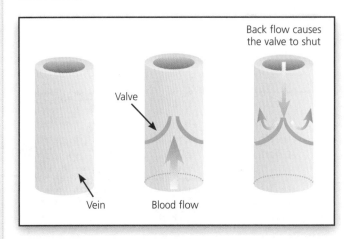

If the valves in veins in the legs were to fail, varicose veins could form. This means that the blood does not move to where it should and it instead drops to the next working valve, causing inflammation and clots.

Blood

The blood transports glucose and oxygen to muscles and other cells. This is because all cells respire and need a supply of dissolved food and oxygen for this to take place. The blood then carries away from the cells waste products such as carbon dioxide (taking it back to the lungs so that it can leave the body).

Transport of Substances in Blood to and from a Muscle Cell

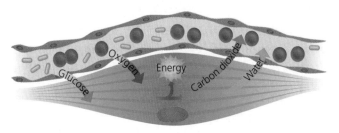

Components of Blood

Blood is a mixture of different components:

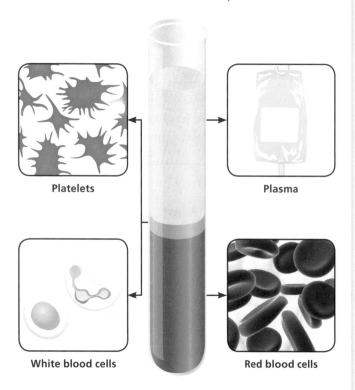

Platelets

Plasma

White blood cells

Red blood cells

Red blood cells transport oxygen from the lungs to the body. They have no nucleus which means they can be packed full with the red pigment haemoglobin, which binds to oxygen to form oxyhaemoglobin.

> **HT** Their biconcave shape provides a larger surface through which oxygen can diffuse.

White blood cells have a nucleus and come in a variety of shapes. They defend the body against microorganisms; some white blood cells engulf and kill microorganisms while others produce antibodies to attack microorganisms (see page 12).

Platelets are tiny particles found in blood plasma. They are not cells and do not have a nucleus. When a blood vessel is damaged, platelets are triggered to clump together to form a meshwork of fibres in order to form a clot and prevent blood leaving the body.

Plasma is a straw-coloured liquid that makes up around 55% of the total blood volume.

It transports:
- nutrients (such as glucose and amino acids)
- antibodies (which defend the body against infection)
- hormones (such as insulin and oestrogen)
- waste (such as carbon dioxide from respiration and urea from the liver).

HT ## Capillary Beds

Blood flows at high pressure from the heart in the artery. The blood reaches its destination via arterioles that branch off the artery and into capillary beds that surround cells.

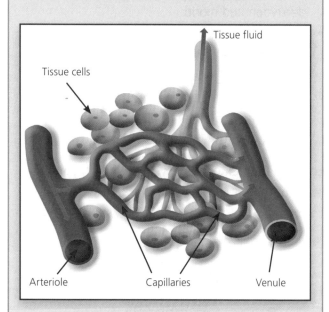

Tissue fluid

Tissue cells

Arteriole   Capillaries   Venule

The flow of blood through a capillary bed is very slow. The plasma leaves the blood and becomes **tissue fluid**. Tissue fluid enables the nutrients required by the cells (e.g. glucose for respiration, oxygen and hormones) to diffuse into the tissue cells.

The tissue fluid also collects and carries away some cellular waste products, such as carbon dioxide and urea. About 90% of the tissue fluid returns to the capillary bed, where it again becomes plasma. It leaves the capillary bed via venules and from there goes to the veins to continue its journey through the body.

Homeostasis

Homeostasis is the maintenance of a constant internal environment. It is achieved by balancing bodily inputs and outputs, while removing waste products. The body has automatic control systems, which ensure that the correct, steady levels of temperature and water are maintained.

To maintain a constant body temperature, the heat gained by the body (including the heat that is released from respiration) has to be balanced with the heat that is lost.

Controlling body temperature requires:

- temperature receptors in the skin to detect the external temperature
- temperature receptors in the brain to measure the temperature of the blood
- effectors (sweat glands and muscles), which carry out the response
- the brain, which acts as a processing centre to receive information from the temperature receptors and to send signals to trigger the effectors.

HT The part of the brain that detects temperature changes in the blood, as well as processing and coordinating the response, is the **hypothalamus**.

Cerebrum

Hypothalamus

Pituitary gland

Cerebellum

If the temperature of the body is too high then heat needs to be transferred to the environment. This is achieved through sweating, since evaporation from the skin releases more heat from the body.

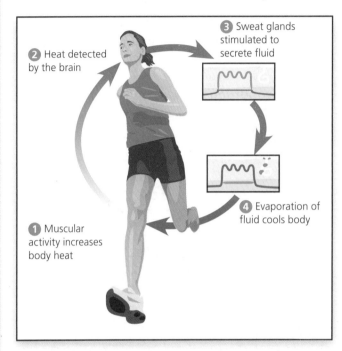

② Heat detected by the brain

③ Sweat glands stimulated to secrete fluid

④ Evaporation of fluid cools body

① Muscular activity increases body heat

If the temperature of the body is too low then the body starts to shiver. Shivering is the rapid contraction and relaxation of muscles. These contractions require energy from increased respiration and heat is released as a by-product, warming surrounding tissue.

Heat Stroke

Heat stroke is the result of an uncontrolled increase in body temperature, i.e. the body cannot lose heat fast enough. The core body temperature is 37°C. If it increases to 40°C this is life threatening. At 41°C the brain stops functioning properly, so it cannot trigger the effectors that would normally lead to heat loss.

Common causes of heat stroke include exercising in very warm or humid conditions, and **dehydration**.

Dehydration can be caused by increased sweating due to very hot temperatures and/or excessive exercise. Dehydration stops sweating from taking place, which leads to the core body temperature increasing even further. If the body is not cooled down then death will rapidly occur.

Vasodilation and Vasoconstriction

Vasodilation is the widening, and **vasoconstriction** is the narrowing, of the blood vessels (capillaries) that run very close to the surface of the skin.

In **hot conditions**, blood vessels in the skin dilate causing greater heat loss, i.e. more heat is lost from the surface of the skin by radiation.

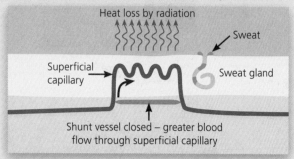

In **cold conditions**, blood vessels in the skin constrict to reduce heat loss, i.e. less heat is lost from the surface of the skin by radiation.

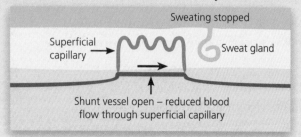

The control of temperature is an example of effectors working antagonistically, which means that a more sensitive, controlled response takes place.

Food and Glucose

Glucose is needed for respiration. When we eat foods containing carbohydrates, enzymes are needed to break them down into monomers. For example, the starch polymer is broken down into glucose monomers by the enzyme amylase.

Processed foods, as compared to fresh foods, are foods that have been changed from their naturally occurring state to make them healthier and / or for convenience. For example, milk is a processed food because it is pasteurised to kill bacteria that could otherwise cause infection.

Many processed foods have additives to improve shelf life, to make them more attractive and to make them taste better. Often sugar is added to the food in high levels, typically in the form of **sucrose**.

Sucrose is a dimer of glucose and fructose monomers joined together.

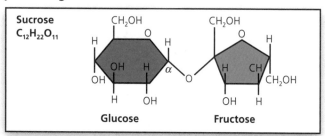

Sucrose is easily broken down in the small intestine into glucose and fructose. Fructose travels to the liver and is metabolised there, while glucose can then travel immediately in the blood. This causes an immediate increase in blood sugar level (a 'sugar rush'), which can cause problems for the body.

Insulin

Insulin is a hormone produced by the pancreas. The presence of insulin causes cells to take in glucose from the blood. It effectively unlocks the cell so that glucose can enter; without the presence of insulin glucose cannot pass into the cell.

Inside the liver and muscle cells the glucose, if it is not immediately used in respiration, is converted into a storage carbohydrate called **glycogen**. Glycogen is a polymer made up of glucose monomers.

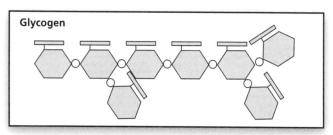

When glucose is absent, the body starts using fat as a source of energy instead.

The body alters the amount of insulin produced, depending on the amount of glucose in the blood.

Diabetes

Diabetes is a condition that develops when glucose can no longer enter a cell to be used in respiration. This can be caused by a variety of reasons. There are two types of diabetes, which are caused in different ways.

Type 1 diabetes typically develops in young individuals up to the age of 40. It is where the insulin-producing cells in the pancreas are destroyed so the pancreas can no longer produce insulin.

The reason why this happens is not yet known. However, sometimes it is triggered by a virus or other infection. To treat type 1 diabetes, sufferers have to monitor blood sugar levels and inject insulin.

Type 2 diabetes is also known as late-onset diabetes. This is where the body does not produce enough insulin or the cells do not respond in the right way to insulin. As a result, the cells in the body are not 'unlocked' to allow the glucose in.

Type 2 diabetes usually affects people over the age of 40, although it has been known to start as early as 25 in people from some cultural backgrounds.

There has been a large increase in people getting type 2 diabetes over the past few decades. This is believed to be due to a poor diet (high in processed foods) or obesity. Treating type 2 diabetes requires a controlled diet and exercise.

Maintaining Constant Blood Sugar Levels

Following a diet in which sugar is released slowly from food is a way of preventing or controlling type 2 diabetes. This is achieved by eating foods with a low **glycemic index** (GI).

The glycemic index is a measure of the complexity of the carbohydrate inside the food. The more complex the carbohydrate, the lower the GI. If the carbohydrate is more complex then it will take longer to digest and break down into sugars, thereby preventing a sugar rush.

Low GI foods include wholemeal bread, fruit and yoghurt.

Medium GI foods include basmati rice and table sugar.

High GI foods include white bread, cornflakes and potatoes.

The table shows the number of deaths and their cause in 250 individuals:

Diet and Exercise History	Deaths due to Heart Disease	Deaths due to Cancer	Accidental Deaths
High GI diet No exercise	50	29	11
High GI diet Moderate exercise	43	20	9
Low GI diet No exercise	6	10	12
Low GI Diet moderate exercise	2	5	10
Diet unknown Exercise unknown	20	11	12

The table suggests that having a poor diet increases your chance of getting heart disease or cancer. In addition, not doing exercise also increases the chances. Interpreting data such as this is an important way of identifying correlations and cause.

Dietary Fibre

Eating foods high in fibre also helps to reduce blood sugar levels. Studies have shown that a high-fibre diet can reduce blood-sugar content by nearly 10%.

The Importance of Exercise

Exercise, along with a healthy diet (high in fibre and complex carbohydrates), helps to prevent cardiovascular disease and maintain a healthy body mass. When the body exercises, fat reserves are broken down and muscle is built up. This helps to keep the BMI in the correct range.

B7 Further Biology (Learning from Ecosystems)

Module B7 (Learning from Ecosystems)

All life on Earth is connected. Changes in one part will impact on another. This topic looks at:

- how ecosystems can be viewed as being open or closed
- how nutrient cycling takes place
- why animals and plants tend to produce large numbers of gametes and offspring
- how ecosystems develop
- how toxins can accumulate in a food chain
- the human impact on the environment.

Closed Loop Systems

An ecosystem relies on inputs to ensure that growth can continue. If **all** of the inputs are met by **most** of the outputs from the ecosystem (i.e. no new inputs are needed), then it is said to be a **closed loop system**.

This means that the waste from one part of the system is used by another part of the system. This is very efficient and means that, undisturbed, an ecosystem can remain that way for an indefinite period. A **perfect closed loop system** would mean that no waste was ever lost and the system could last infinitely.

In reality, it is impossible to have a perfect closed loop system in an ecosystem. This is because organisms migrate out of the area and nutrients can be lost because they are transferred to another ecosystem by wind or water.

However, to be a **stable** ecosystem the outputs must be balanced by gains – a rainforest, such as the Amazon, is an example of this.

'Waste' Products

The 'waste' products in an ecosystem are:

- **oxygen** (from photosynthesis)
- **carbon dioxide** (from respiration)
- **dead organic matter** (fallen leaves, petals, fruit, faeces, the remains of dead plants and animals, etc.).

Although classed as being 'waste', it is only so for the organism that produced it. Other organisms may find the waste extremely useful, e.g. humans need oxygen (waste product of photosynthesis) and many drink alcohol (waste product from yeast).

Using 'Waste'

The chemicals that make up 'waste' materials can be used either as a food or as reactants for different chemical reactions in the organism (i.e. plants, animals or microorganisms).

To break down food products, organisms often have to use **digestive enzymes**. Digestive enzymes break down large molecules into sizes that are small enough to pass through the intestine, so that they can be used to make new larger molecules.

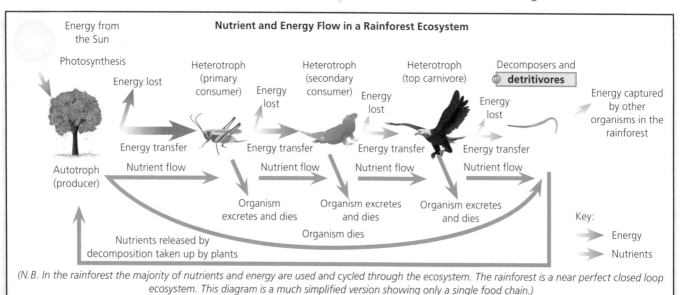

Nutrient and Energy Flow in a Rainforest Ecosystem

Energy from the Sun

Photosynthesis

Energy lost

Heterotroph (primary consumer)

Energy lost

Heterotroph (secondary consumer)

Energy lost

Heterotroph (top carnivore)

Energy lost

Decomposers and **detritivores**

Energy captured by other organisms in the rainforest

Energy transfer — Nutrient flow

Autotroph (producer)

Organism excretes and dies

Organism excretes and dies

Organism excretes and dies

Organism dies

Key:
— Energy
— Nutrients

Nutrients released by decomposition taken up by plants

(N.B. In the rainforest the majority of nutrients and energy are used and cycled through the ecosystem. The rainforest is a near perfect closed loop ecosystem. This diagram is a much simplified version showing only a single food chain.)

Digestive Enzymes

If it were not for microorganisms, waste in the environment would build up to intolerable levels. Microorganisms use enzymes to break down the different food groups.

Breakdown of Protein
Proteins are broken down to amino acids by **proteases**.

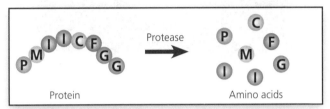

Breakdown of Lipids
Lipids (oils and fats) are broken down into fatty acids and glycerol by **lipases**.

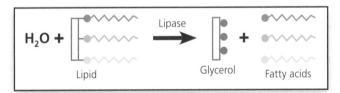

Breakdown of Carbohydrates
Carbohydrates are broken down into sugars by **carbohydrases** (e.g. amylase).

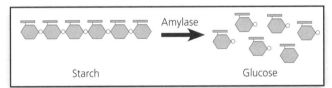

As microorganisms can grow very rapidly, waste can be broken down very quickly.

Closed Loop System Diagrams

Closed loop system diagrams can be used to interpret how resources cycle through the system and to help biologists understand how all the parts of the ecosystem work together. Questions that can be answered include: Are there any unexpected losses? What might be causing the system to become open?

In the early 1990s a project took place to investigate closed loop systems. Biosphere 2 was a closed system where an artificial habitat was set up for humans to live in. Everything inside a dome was isolated from the outside world. The results were analysed and published in a number of scientific journals.

Interpreting the Data in the Diagram
There was a big drop in oxygen levels over the 16-month experiment. The number of plants dropped – this would have helped reduce the amount of oxygen as fewer plants photosynthesised. This drop in the number of plants may have been due to the decrease in pollinating insects, which would affect plant reproduction. Insect pests would have also contributed to the drop in the number of plants.

The amount of water decreased, whilst the amount of carbon increased. This suggests that there must have been a breach in the dome, allowing water to escape and extra carbon (in living things) to enter.

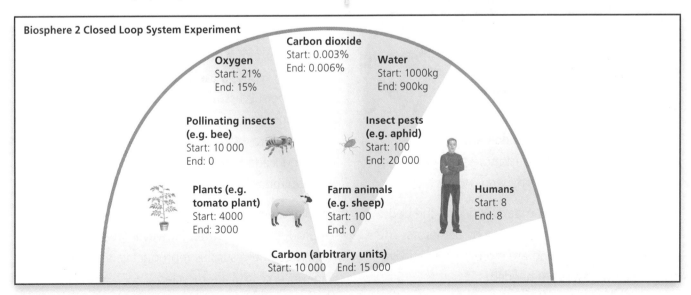

Biosphere 2 Closed Loop System Experiment

Reproduction

Animals

To reproduce successfully, organisms need to use a strategy that maximises the chances of the offspring reaching adulthood and reproducing themselves.

In most ecosystems there is competition between species. This means organisms have to invest energy and resources in achieving their goal of reproducing.

Females usually produce large numbers of eggs, while males produce large quantities of sperm. This ensures a high chance that at least one successful fertilisation will occur.

Growing offspring are a good source of food, so there is always a high chance that some of the offspring will be eaten by predators. Producing large numbers of offspring helps to mitigate against this loss.

With organisms such as frogs, the embryos develop together in a mass of frog spawn. This ensures that, even if some of the offspring are eaten, others will survive.

Frog Spawn

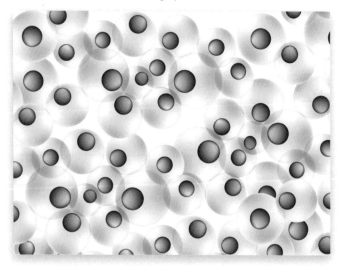

Where the numbers of offspring are lower, the parents usually have to invest more time in staying with and protecting the offspring.

In mammals the offspring develop inside the body, so they are less likely to be eaten by predators. This means that the offspring are more likely to survive and therefore fewer are needed.

An Ultrasound Scan of a Human Embryo

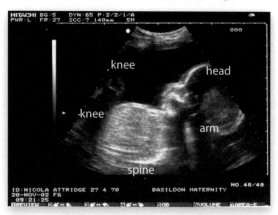

Plants

Plants, which cannot move around, also have to employ strategies to ensure the success of the species. These strategies involve the following:

- **Flowers** – to attract insects in sufficient numbers to ensure that reproduction occurs, large numbers of attractive-looking and nice-smelling flowers have to be produced. Not all flowers will be successfully fertilised.
- **Pollen** – large quantities of pollen have to be produced. This is taken either by the wind (in wind-pollinated plants) or carried by insects (in insect-pollinated plants) to the destination plant. There is no guarantee that the pollen will land where it is supposed to.
- **Fruit** – having been fertilised, the next stage in a flower's life cycle is to produce fruit. A fruit contains the seed and is grown to encourage animals to take the fruit and eat. Seeds, which are quite resilient, pass through the digestive system unharmed and are deposited in faeces. The faeces is a good source of nutrients (fertiliser), which will help the seed to grow. Not all seeds will be successful – they may be deposited in the wrong area, or chewed up and damaged. Therefore, large numbers of fruit need to be produced to ensure that some of the seeds grow to adulthood, so that they can reproduce themselves.

It may seem wasteful producing large numbers of reproductive structures. However, in a stable ecosystem, the surplus reproductive structures that are unsuccessful are recycled.

Plants and Soil

As well as contributing to the inputs and outputs of an ecosystem, organisms (particularly plants) have a big physical effect on the ecosystem itself. Plants have roots, which effectively bind the soil together. The larger the plant, the more extensive its root system. This means that the soil is bound together more effectively and is unlikely to be washed away.

The foliage of plants also reduces the effect of heavy rainfall washing away the surface soil. Most of the rainfall hits leaves, so water only gets to the surface gently and indirectly.

Vegetation can also alter the climate. It can prevent extremes of temperature by reducing the amount of carbon dioxide (a greenhouse gas) in the atmosphere. Rainforests transport a high quantity of water via transpiration. This means water is cycled from the ground to the air and this can stimulate cloud formation.

Humans and Ecosystems

Humans benefit from and depend on ecosystems to provide a huge range of resources and processes. These are known as **ecosystem services**.

In 2004 the United Nations stated that humans are dependent upon ecosystem services and defined what a healthy ecosystem requires. These include providing clean air, water, soil, mineral nutrients, pollination and the presence of fish and game. If all of these are present, then the ecosystem is healthy.

Human Activity – Effects

Systems involving humans are not closed loop systems. Human waste from households, agriculture and industry leaves the system as non-recycled waste, as well as through pollution from burning fossil fuels. This means the system is losing resources. Sometimes the waste can build up to harmful levels, which then affect other organisms.

Human activities can unbalance an ecosystem, changing the inputs and outputs so much that the ecosystem can no longer adapt. This means that the system is no longer a closed loop.

Households produce waste (e.g. unconsumed food, packaging, etc.), which is an additional input to the ecosystem, changing its balance.

Industry produces large amounts of waste and pollution, which alter the environment. Carbon dioxide, although needed for photosynthesis, causes climate change through the process of enhanced global warming. Pollution by chemical waste can kill parts of food webs.

Fossil fuels are burned by homes, vehicles and industry, producing levels of carbon dioxide and other gases in much higher quantities than occur naturally.

In small enough quantities, the waste caused by human activity could be accommodated by the environment. However, with a growing population and industrial expansion, chemical waste has built up to harmful levels.

Plants and Soil

Rainfall can hit the surface easily

Rain does not directly hit the forest floor

Root system shallow – soil at risk of being washed away

Extensive, deep, root system – soil stable

⒣ Bioaccumulation

Bioaccumulation is where a chemical that is deliberately or accidentally introduced into the environment starts to build up in concentration at each level in a food chain.

For example:

Grass	Rabbit	Fox
1 plant	Eats 100 grass plants	Eats 10 rabbits
1 unit	100 units	1000 units

In Ontario, Canada, in 1970, mercury waste was released into nearby lakes from factories that were making batteries.

Microorganisms in the sediment converted the mercury into a compound called methyl mercury. This then entered the food chain via phytoplankton.

These were then eaten by small fish, which were in turn eaten by other fish. As each organism ate a lot of the organisms in the previous trophic level, the concentration of methyl mercury increased.

Eventually fishermen caught and sold the fish. Customers therefore consumed dangerously-high levels of methyl mercury.

Consuming high levels of methyl mercury can cause speech problems, poor balance, heart attacks, blindness and birth defects.

Key: ● = Presence of toxin (size indicates relative concentration)

Human Activity – Removing Resources

When humans take away too many resources as biomass, then this reduces the amount available to be recycled within the ecosystem.

For example, cutting down a rainforest for wood removes a large number of trees. However, the tree canopies would have protected the soil from rainfall and the roots would have bound the soil together. The trees would have also provided habitats for other organisms.

Removing too many trees causes the closed ecosystem to become open. The soil dries out and is blown away, and the organisms that relied on the trees for survival die. The entire ecosystem is damaged to the point of destruction.

Another example is over-fishing. Removing a few fish enables the ecosystem to adapt, but taking out too many disrupts the food chain and causes irreversible changes.

Biologists have carried out studies and have developed an explanation for what happens when humans take too many resources out of an ecosystem. This reduces the amount available to be recycled.

Removing Vegetation

Human activities have changed the landscape since the Stone Age. When the population was small, the environment was able to adapt.

During the time of Henry VIII (1491–1547) mass deforestation occurred in England. Wood was removed to build the warships in his navy. As a result, the majority of England is no longer covered by forest.

An increasing population requires more food, so farming has replaced the natural vegetation with agricultural crops and livestock. The consequences of this land-use change include a loss of biodiversity (especially if the farming is a crop monoculture).

The removal of trees and shrubs to make way for livestock or crop plants can reduce the soil quality as it is more likely to be washed away by rain. Overgrazing by farm animals also removes the remaining plants and their roots, which bind the soil together.

When washed away, the soil ends up in nearby rivers and streams, silting them up and altering the flow of the water. If the farming is not managed properly (with crop rotation), then ultimately what used to be forest can end up as desert. This process is called **desertification**.

The removal of natural resources is only sustainable if they are used at a rate at which they can be replaced. To try to ensure farming and fishing are sustainable, quotas and requirements for restocking or replanting are often imposed to allow ecosystems to recover from harvesting and to prevent over-fishing.

ⓗⓣ Eutrophication

Eutrophication is where an excess of nutrients is put into a system, causing the productivity of the system to increase. Unfortunately, this causes the balance of organisms to change, often drastically and irreversibly.

An example is a farmer who has applied too much fertiliser (organic or chemical) to a crop. The excess fertiliser is washed away into nearby rivers and streams.

The fertiliser, which is high in nutrients, causes a large increase in algae. These rapidly choke the watercourse. The algae then die. Bacteria, which decompose the algae, reproduce in very large numbers, using up the available oxygen in the water. This then causes animals, such as fish, to die.

Eventually the watercourse is unable to support life, other than that which can survive on putrefying plant and animal material.

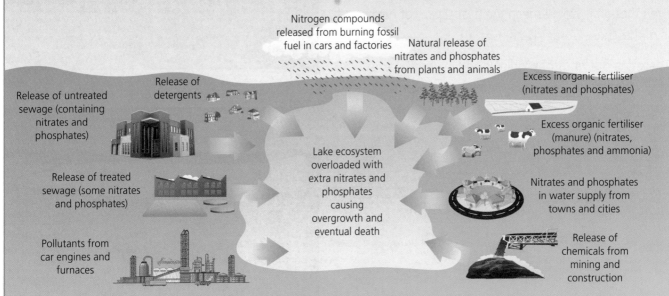

Nitrogen compounds released from burning fossil fuel in cars and factories

Natural release of nitrates and phosphates from plants and animals

Release of detergents

Excess inorganic fertiliser (nitrates and phosphates)

Release of untreated sewage (containing nitrates and phosphates)

Excess organic fertiliser (manure) (nitrates, phosphates and ammonia)

Lake ecosystem overloaded with extra nitrates and phosphates causing overgrowth and eventual death

Release of treated sewage (some nitrates and phosphates)

Nitrates and phosphates in water supply from towns and cities

Pollutants from car engines and furnaces

Release of chemicals from mining and construction

Crude Oil

Crude oil is formed from the remains of plants and animals that died millions of years ago. The biomass is covered by silt and rock, and subjected to immense pressure and heat. Over millions of years, this causes the biomass to be converted into oil.

The Sun originally supplied the energy for the plants and animals to grow. In effect, all the energy that was stored in each plant and animal is now compressed in a much smaller space, in oil. When we burn oil, we are effectively using fossil sunlight energy.

Due to the immense length of time needed to create it, crude oil is regarded as being non-renewable and is **not** part of a closed loop system.

Could We, Should We?

Sunlight is a sustainable source of energy – it will not run out for another five billion years – and it allows sustainable agriculture and the growth of natural ecosystems.

Nowadays, when we understand more about how ecosystems work and how they can fail, we need to balance the need to conserve natural ecosystems with the needs of the human population.

Would it be right to let a community starve just to protect an ecosystem? Should a river be dammed to provide water and energy for an increasing population, even though it would mean the destruction of habitats (e.g. the Three Gorges Dam, China)?

These are the kinds of problems and issues that exist, and scientists are expected to help to answer them.

Creation of Crude Oil

Marine plants and animals die and sink to the bottom of the seabed

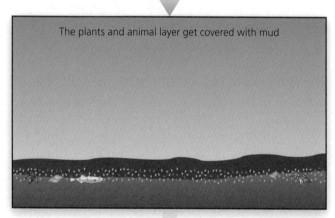

The plants and animal layer get covered with mud

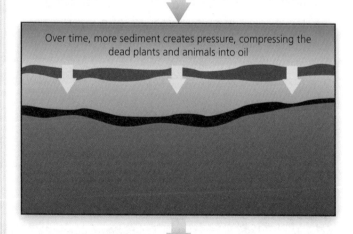

Over time, more sediment creates pressure, compressing the dead plants and animals into oil

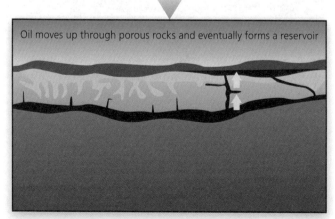

Oil moves up through porous rocks and eventually forms a reservoir

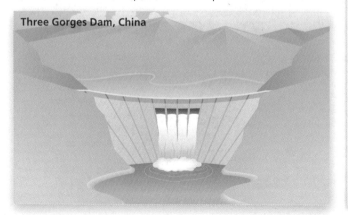

Three Gorges Dam, China

Further Biology (New Technologies)

The understanding of biology has enabled the introduction of a variety of new technologies that are already having, or will have in the future, an impact on our daily lives. This topic looks at:

- how microorganisms can be used to produce resources
- genetic modification to alter products
- how DNA profiling works
- nanotechnology – what it is and the potential benefits
- stem cells as a medical treatment.

Bacteria

Bacteria are ideally suited for industrial and genetic processes. This is due to:

- rapid reproduction – bacteria can double in number every 20 minutes
- the presence of plasmids – circular DNA molecules that can be transferred easily between bacteria
- simple biochemistry – which makes the bacteria easier to understand and alter
- their ability to make complex molecules – bacteria can produce molecules that can be used medicinally
- a lack of ethical concerns – when using bacteria, people do not regard them as the same as higher-order animals like rats and chimpanzees.

Industrial Fermentation

Bacteria and fungi can be grown on a small scale in test tubes. By maintaining the ideal conditions for a microorganism, it is possible to expand the process to an industrial (very large) scale using the process of **fermentation** (see page 40) and a **fermenter**.

A fermenter is a controlled environment that provides ideal conditions for microorganisms to live in, feed and produce the proteins needed.

A fermenter allows the continuous culture of large quantities of microorganisms or their products, namely:

- antibiotic and medicinal products – it is possible to extract specific antibiotics and vaccines from bacteria and fungi grown in a fermenter
- single-cell proteins / hormones – insulin is a hormone that can be produced via bacteria in a fermenter
- enzymes for commercial products – for example in biological washing powders that wash at 30°C and biofuels (such as ethanol)
- enzymes for food processing – for example chymosin, a vegetarian substitute for rennet, which is used to make cheese. Rennet is usually extracted from the stomach of animals. Tofu is another example of a food made using enzymes.

An Industrial Fermenter

1. **Stirrer** – keeps the microorganisms in suspension and maintains an even temperature
2. **Water-cooled jacket** – removes heat produced by the respiring microorganisms
3. **pH probe** – monitors the pH of the mixture
4. **Temperature probe** – monitors the temperature of the mixture
5. **Outlet tap** – collects the proteins
6. **Sterile air supply** – provides oxygen for respiration. Air is sterilised to prevent contamination

Genetic Modification

DNA is the genetic material of all organisms and it contains the genes that code for the particular proteins an organism needs. Proteins produced by one organism may not necessarily be produced by another.

By carrying out genetic modification, the gene that produces a desirable protein can be inserted into another organism so that it also produces the required protein.

Once transferred into the target organism, the gene works and the target organism produces the new protein.

These are the steps:

1. The target gene is selected and isolated.
2. The gene is now replicated (copied exactly and increased in number).
3. The gene is inserted into the target bacterium by a **vector**. This could be a **virus** or a **plasmid** (a double-stranded DNA molecule that replicates independently of the bacterial chromosomal DNA). Viruses spread easily into cells. The virus incorporates the DNA with the target gene into the bacterium. The bacterium will now activate the gene and produce the protein. When using a plasmid, the plasmid is cut open and the gene inserted. The plasmids spread among the population of bacteria. In addition, the plasmids are copied when the bacteria reproduce. This is the main way in which the plasmid increases in number.
4. The modified individuals are selected. It is usual to insert a gene for antibiotic resistance as well as the target gene. The specific antibiotic is applied to a plate of bacteria. Those bacteria with antibiotic resistance will also carry the desired gene and survive, while those that do not have antibiotic resistance will not have the desired gene and will therefore die. The surviving bacteria can then be harvested and grown in a fermenter.

Using Genetic Modification

Genetic modification is now used to create an increasing number of drugs and hormones to treat patients.

For example, genetically-modified microorganisms are used in the production of insulin for people who have diabetes (see page 71). Millions of people suffer from this condition and require injections of insulin. Until recently, insulin was taken from pigs and cows. However, it is now possible to use genetically-modified bacteria to produce human insulin.

Genetic modification can also be used when growing crops. Farmers often have problems with weeds. By creating crops with resistance to a herbicide, the farmer can use that herbicide to kill weeds without destroying the crop.

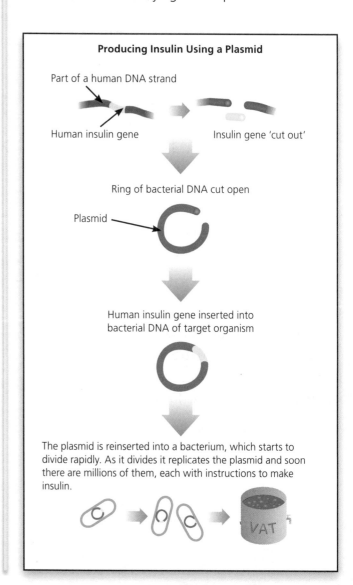

Producing Insulin Using a Plasmid

Part of a human DNA strand

Human insulin gene

Insulin gene 'cut out'

Ring of bacterial DNA cut open

Plasmid

Human insulin gene inserted into bacterial DNA of target organism

The plasmid is reinserted into a bacterium, which starts to divide rapidly. As it divides it replicates the plasmid and soon there are millions of them, each with instructions to make insulin.

VAT

Genetic Testing

The vast majority of DNA is identical in all humans. In fact, humans share 99.9% of their DNA with chimpanzees. Identifying specific individuals, rather than a species, requires a more specific technique – **DNA profiling**.

DNA profiling was invented by Sir Alec Jeffreys in 1985. It involves using specific areas of DNA that repeat regularly, called **microsatellites**.

Although variable, microsatellites are passed onto children. Everyone's microsatellites are different, a little like a fingerprint. This means that DNA profiling can be used to identify the paternity of a child or to identify the perpetrator of a crime.

DNA testing can also look for alleles of specific genetic disorders. This enables people to find out the chances of getting a genetic disorder and take action in advance.

Genetic testing involves extracting the DNA from cells. In forensic science, these can be from any type of cell (e.g. cheek, skin, hair, etc.). In a sample of blood, DNA can be extracted and isolated from white blood cells.

Enzymes and detergents are used to release the DNA. Once the DNA has been washed and non-DNA material removed, it is subjected to a process called **polymerase chain reaction (PCR)**.

PCR is used to **amplify** the sections of the genome that are being investigated (i.e. the microsatellites for identity or the gene alleles for disorders).

PCR works by using the original DNA as a template and adding new copies using the enzyme, DNA polymerase, and **primers** (building blocks of DNA).

PCR is a chain reaction because, once started, the DNA exponentially increases in quantity, so that in a short period of time there is plenty of identical DNA for the geneticists to work with.

Humans Share Nearly All Their DNA With Chimpanzees

Steps in the Polymerase Chain Reaction

Genomic DNA

Target sequence

DNA denatured (broken) and new bases added to form new strands

Primers

Cycle 1 yields two molecules

New nucleotides

Cycle 2 yields four molecules

Cycle 3 yields eight molecules

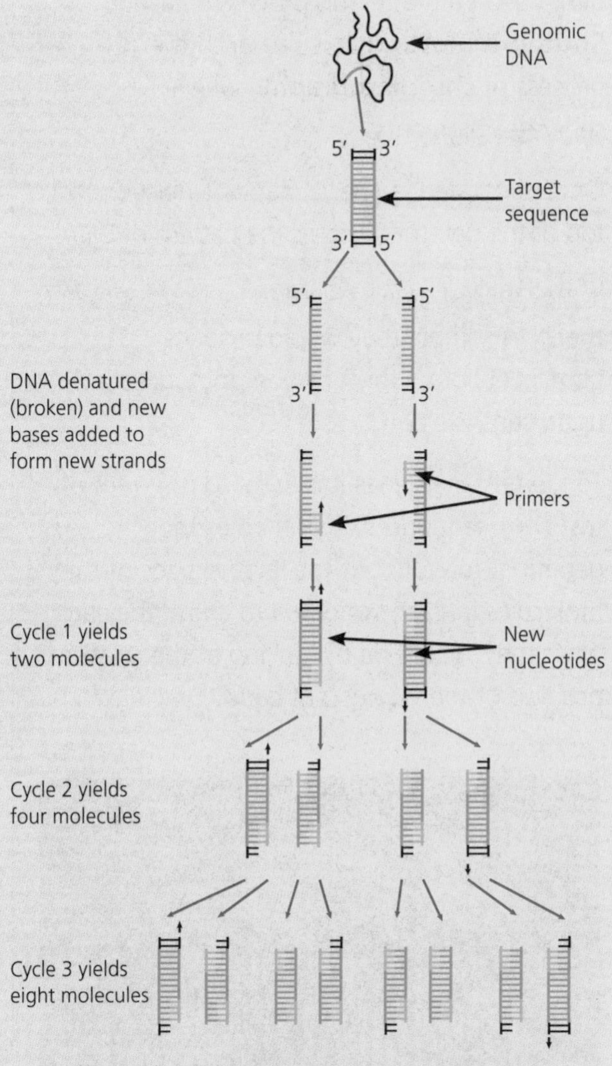

This process continues until many copies of the target sequence are created

Genetic Testing (Cont.)

Gene probes (markers) are created that are mirror copies of the target allele or microsatellite region. The gene probes are attached to a fluorescent chemical that emits ultraviolet light. If the target segment of DNA is present in the DNA sample, the gene probe will attach to it.

The samples are put into the wells of a polyacrylamide gel and an electric current is applied. The DNA fragments in the sample move down the gel at a rate determined by their charge and mass. This means that the genes separate out. This process is called **Southern blotting**. It is similar to the process of **chromatography**, which separates pigments.

Once separated, a special paper is placed on top of the gel. This absorbs the pattern of DNA.

If the labelled genes or alleles were present in the DNA sample they would produce ultraviolet light, which can be photographed using sensitive film.

To identify the genes it is usual to run a set of test DNA sequences. This enables the geneticist to work out the relative sizes of the molecules in each band and to check they are what they should be by looking at the relative position of the fluorescent band.

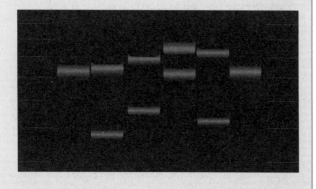

Nanotechnology

Nanotechnology promises to revolutionise biology. It is the application of matter on an atomic scale, i.e. extremely small (between 1 and 100nm in at least one dimension). This is the same size as some molecules.

To compare:
- 1mm = 1000µm
- 1µm = 1000nm

So, there are 1 000 000nm in a mm.

In biology, the applications that are being developed include the following:
- Nanoparticles – these can deliver chemotherapy drugs directly to cancer cells.
- Silver nanoparticles – silver has antimicrobial properties. It would be extremely expensive coating objects with silver. However, it is a lot cheaper using silver nanoparticles impregnated into different materials. Deodorants and bandages have been developed with silver added to them – the silver helps to kill microorganisms.
- Tissue engineering – this is where scientists are looking to alter biological processes on an atomic scale, i.e. manipulating individual atoms or creating nanomachines to carry out processes.

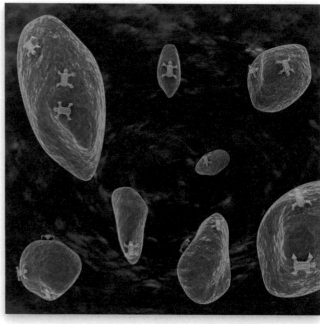

In the future it may be possible to engineer nanorobots that can repair cells in the body.

Nanotechnology and Food

Nanotechnology is being applied to the food industry. Wasted food is a big problem; it may end up in landfill. Food has to be sold with a use-by date telling the customer when it should be eaten. If the food is eaten after the given date it may have spoiled, i.e. bacteria may have increased in number to make it unsafe to eat.

The use-by date is a scientific estimate, based on experiments. Samples of the food in question are placed on an agar plate. The plate is left and checked at intervals to see when the bacteria reach unsafe levels. However, as a precaution and because each batch of food varies, the use-by date is earlier than it could be.

Nanosensors are now being developed that can be incorporated into packaging. The sensors change colour when they detect the gases that are produced when food goes off.

Clear packaging has turned red, indicating that the peppers have gone off.

Oxygen is a very reactive molecule that can cause rapid food deterioration. Manufacturers can prevent this by filling food bags with inert nitrogen gas. In addition, nanoparticles can be incorporated into packaging in order to prevent oxygen entering, improving the shelf life of the food product.

Nanoparticles have also been added to the plastic used to make fizzy drink bottles. This prevents carbon dioxide from leaking out, again improving the shelf life.

Methods such as those described above should help to reduce the amount of food that is thrown away.

Stem Cell Technology

Stem cells are being used to reverse damage to the body.

Leukaemia

Stem cells can help to treat leukaemia, a disease that kills white blood cells. Blood cells are made from the body's own adult stem cells in the bone marrow (a spongy material in the centre of some bones). Traditionally, a leukaemia patient would need to have their own bone marrow removed and replaced with that from a tissue-matched donor.

However, using stem cells that have been harvested (and, if necessary, genetically manipulated) from the patient's own body has a significant advantage. It means that the patient has a new complement of blood cells that are genetically the same as him / her. This reduces the need for a bone marrow transplant, which brings with it the risk of rejection.

Spinal Injuries

If a patient survives a broken spine, they will be paralysed from the break downwards. Spinal nerves do not re-grow, so this is a permanent and devastating injury. Stem cells are being investigated as they may allow the nerves to be reconnected and enable paralysed patients to move again.

Biomedical Engineering

Scientists and engineers are becoming more efficient at designing and creating replacement organs and body parts. This includes replacing faulty heart valves and designing pacemakers that keep the heart beating when the heart's own pacemaker (the **sinoatrial node**, a special tissue that causes the synchronisation of beating muscle) fails.

Replacement parts such as these have to be designed exceptionally well as they have to work perfectly on a daily basis over many years and decades.

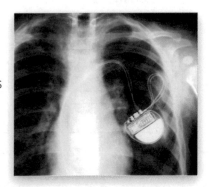

Exam Practice Questions

1 This question is about the circulatory system.

(a) Draw straight lines to join the parts of the circulatory system to their correct descriptions. **[3]**

Part	Description
Aorta	Transports oxygenated blood to the heart from the lungs
Ventricle	An upper chamber of the heart
Atrium	A vein that returns deoxygenated blood to the heart from the body
Vena cava	Transports oxygenated blood at high pressure to the body
Pulmonary artery	Transports deoxygenated blood to the lungs
Pulmonary vein	A lower chamber of the heart

(b) List the four components of blood. **[4]**

2 Ethan is 1.79m tall and weighs 91kg. What is his BMI and what does it mean? **[6]**
   ✐ *The quality of written communication will be assessed in your answer to this question.*

3 Nanotechnology could deliver improvements to our lives. However, some people say that it is 'not worth the risk'. Explain why we cannot remove all risk and what policy-makers have to consider before going ahead. **[6]**

   ✐ *The quality of written communication will be assessed in your answer to this question.*

4 The following steps refer to the formation of crude oil. They are in the wrong order. Put letters in the empty boxes to show the correct order. **[1]**
   A Over time, more sediment creates pressure, compressing the dead plants and animals into oil.
   B Oil moves up through porous rocks and eventually forms a reservoir.
   C The plants and animal layer get covered with mud.
   D Marine plants and animals die and sink to the bottom of the seabed.

   Start [ ][ ][ ][ ]

5 Draw straight lines to join the correct enzyme to the correct substrate and breakdown product. **[3]**

Substrate	Enzyme	Breakdown Product
Protein	Lipase	Amino acids
Lipids	Cellulase	Sugars
Cellulose	Protease	Fatty acids and glycerol

6 Four students are discussing genetic modification.

Genetically modified bacteria can have their genes altered by inserting new genes into plasmids.

It is morally wrong to hurt bacteria as they may have feelings.

Bacteria are bad – we should wipe them out.

Genetic modification is against religion.

Skirmante   **Samuel**   **April**   **Adam**

Who is putting forward a scientific statement? **[1]**

7 This question is about nanotechnology.

(a) What is the correct range for nanotechnology? Put a ring around the correct answer. **[1]**

1 to 1000cm   **0.1 to 100mm**   **1 to 100nm**   **1 to 1000nm**

(b) Explain what a nanosensor incorporated into packaging could do. **[6]**
✎ *The quality of written communication will be assessed in your answer to this question.*

HT **8** On the diagram below indicate the location of the hypothalamus, cerebrum, cerebellum and pituitary gland. **[2]**

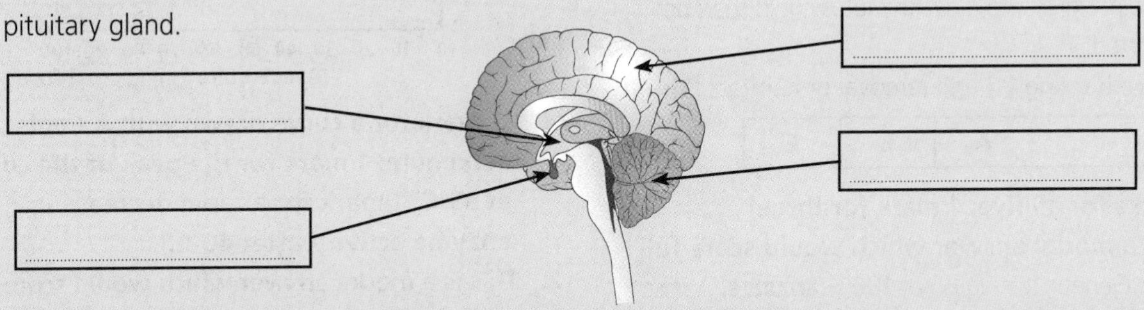

9 DDT is a chemical that was spread onto crops. It resulted in bioaccumulation. Which of the following statements best describes the meaning? Put a tick (✓) in the box next to the correct answer. **[1]**

Bioaccumulation is where the DDT is magnified in crops.

Bioaccumulation is where the concentration of DDT decreases as the trophic level increases.

Bioaccumulation is where the concentration of DDT increases as the trophic level increases.

Bioaccumulation is where DDT gets higher in a food chain.

10 Describe and explain what is meant by the term 'eutrophication'. **[6]**
✎ *The quality of written communication will be assessed in your answer to this question.*

Answers

Unit A161 (Pages 30–31)

1. **(a)** DD **and** Dd **should be ticked. [Both must be correct for 1 mark.]**

   (b)

		♂	
		B	**b**
♀	**b**	Bb	bb
	b	Bb	bb

   [1 mark for the correct pairs of alleles in a column (2 marks in total).]

2. **(a)** Thick mucus in lungs; Chest infections; **and** Breathing difficulties **should be ticked**.

   (b) This is a model answer which would score full marks: Huntington's disease is an incurable genetic disorder that causes suffering. However, people with the condition do not develop it until later life. It would be unethical to abort a fetus carrying the disorder as it could lead a normal life if allowed to develop as normal. Another issue is that the genetic test may be incorrect, giving a false positive, and a healthy fetus could then be aborted.

3. Excessive hunting **[1]** and removal of habitats **[1]**.

4.

D	C	A	B	E

   [2 marks for all five; 1 mark for three]

5. **This is a model answer which would score full marks:** Genetic testing is not 100% accurate. A false positive would lead to a person worrying about a condition they do not have, whilst a false negative would mean that a carrier does not avoid lifestyle factors that could trigger the disorder. Some disorders are incurable and, therefore, knowing about it in advance is not going to help. Companies could also get hold of the results, which may mean a carrier cannot get a job or life insurance.

6.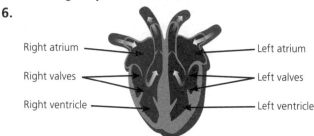

   Right atrium — Left atrium
   Right valves — Left valves
   Right ventricle — Left ventricle

   [1 mark for each correct pair of labels.]

7. Genotype – The alleles present for a gene in an individual

   Phenotype – The characteristics expressed in the environment

   Allele – A version of a gene

   Heterozygous – Possessing one of each allele type

   Homozygous – Possessing two of the same alleles

8. Enough of the population is vaccinated to avoid an epidemic **should be ticked.**

9. 10%

Unit A162 (Pages 58–59)

1.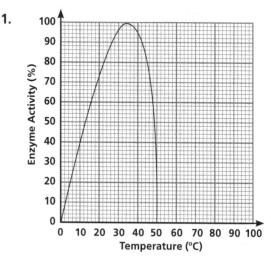

   [1 mark for a curve drawn with a single maximum; 1 mark for the peak of the curve at 35°C; 1 mark for a rapid decrease in enzyme activity after 40°C.]

2. **This is a model answer which would score full marks:** Aerobic respiration involves the reactants glucose and oxygen reacting to produce water and carbon dioxide, with the release of energy. Anaerobic respiration in humans involves glucose being converted to lactic acid, with the release of energy. In plants and microorganisms such as yeast the products of anaerobic respiration are ethanol and carbon dioxide, with energy released. The energy produced by aerobic respiration is much greater than that produced by anaerobic respiration.

3. Water moving from plant cell to plant cell; A pear losing water in a concentrated salt solution; **and** Water moving from blood to body cells **should be ticked**.

4. have a scientific mechanism; be published; **and** be repeatable **should be ticked [2 marks for all three; 1 mark for two].**

5. 16 cell stage / After the 8 cell stage

6. **(a)** The plant will grow to the left / towards the light.

   (b) survival; nervous system / CNS

7. **(a) Any three from:** Stepping; Grasping; Startle; Sucking; Rooting

   (b) This is a model answer which would score full marks: Reflexes enable an organism to react to a stimulus without thinking. The response is quicker than it would be if the organism were to think about how to respond. This may mean the organism survives longer. For example, some simple animals react to changes in light levels, which could indicate a predator was overhead.

   (c) Conditioned reflexes

8. $C_6H_{12}O_6 + 6O_2 \rightarrow 6CO_2 + 6H_2O$
   [1 mark for reactants; 1 mark for products]

9. UUAUCGGCGGUUAAUCCG

10. Ryan

Unit A163 (Pages 84–85)

1. **(a)** Aorta – Transports oxygenated blood at high pressure to the body
   Ventricle – A lower chamber of the heart
   Atrium – An upper chamber of the heart
   Vena cava – A vein that returns deoxygenated blood to the heart from the body
   Pulmonary artery – Transports deoxygenated blood to the lungs
   Pulmonary vein – Transports oxygenated blood to the heart from the lungs

   (b) Plasma, red blood cells, platelets and white blood cells

2. **This is a model answer which would score full marks:** BMI stands for body mass index. To calculate Ethan's BMI you need to take his mass in kilograms and divide it by the square of his height in metres. A healthy BMI is between 18.5 and 25. Ethan's BMI is approximately 28.5, which means he is classed as being overweight. He needs to exercise and eat a healthier, low fat diet.

3. **This is a model answer which would score full marks:** Nothing is without risk – even not doing something carries a risk. Sometimes we do not even know what the risks are until they happen. The consequences of some risks can be greater than others, so policy-makers have to balance the potential risk with the potential benefit. If the benefits are greater and will help many people, then it can go ahead.

4.

| D | C | A | B |

5. Protein – Protease – Amino acids
   Lipids – Lipase – Fatty acids and glycerol
   Cellulose – Cellulase – Sugars

6. Skirmante

7. **(a)** 1 to 100nm **should be ringed.**

   (b) This is a model answer which would score full marks: Nanosensors can be incorporated into packaging so that the packaging changes colour when the gases produced when food starts spoiling are detected. This would help prevent food wastage as well as cases of food poisoning. Nanoparticles could also be added to prevent oxygen entering the food packet. This would improve shelf life. Nanoparticles can be added to drink bottles to stop carbon dioxide leaking out, again improving shelf life.

8.
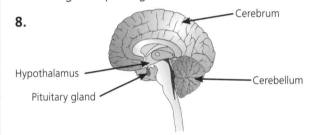

   [2 marks for four; 1 mark for two]

9. Bioaccumulation is where the concentration of DDT increases as the trophic level increases **should be ticked.**

10. **This is a model answer which would score full marks:** Eutrophication is where pollutants (nitrates and phosphates) enter a river or lake. The source of the pollutants can be industry or agriculture (from excessive inorganic or organic fertilisers). The pollutants cause a large increase in algae, which rapidly choke the watercourse. The algae then die. Bacteria, which decompose the algae, reproduce in very large numbers, using up the available oxygen. Eventually all animals die and the river or lake is unable to support life other than that which can survive on rotting plant and animal material.

Glossary

Adaptation – the gradual change of a particular organism over generations to become better suited to its environment.

Aerobic respiration – respiration using oxygen; releases energy and produces carbon dioxide and water.

Allele – an alternative version of a gene.

Anaerobic respiration – respiration that takes place in the absence of oxygen.

Antibody – a protein produced by white blood cells to inactivate disease-causing microorganisms.

Antigen – a marker on a cell surface or on the surface of a disease-causing microorganism.

Antimicrobial – a chemical that kills (or inhibits the growth of) bacteria, fungi and viruses. Antibiotics and antivirals are examples.

Artery – a blood vessel that carries blood away from the heart.

Asexual reproduction – cells divide into two identical daughter cells that are also identical to the parent cell.

Autotroph – an organism that makes its own food.

Biodiversity – the variety of living organisms in an ecosystem.

Biomass – the mass of living matter in a living organism.

Biotechnology – the use of organisms, parts of organisms, or the process the organisms carry out, to produce useful (chemical) substances.

Blood pressure – the force of blood exerted on the inside walls of blood vessels.

BMI (body mass index) – a calculation that compares a person's mass against their height to see if they have a healthy mass.

Bone – rigid connective tissue that makes up the human skeleton.

Capillary – a blood vessel that connects arteries to veins; where the exchange of materials takes place.

Cartilage – smooth connective tissue that covers the ends of bones in a joint.

Cell – the fundamental unit of a living organism.

Central nervous system (CNS) – the brain and spinal cord; allows an organism to react to its surroundings and coordinates its responses.

Chromosome – a coil of DNA made up of genes, found in the nucleus of plant / animal cells.

Clone – a genetically identical offspring of an organism.

Conditioned reflex – a specific reflex response produced when a certain stimulus is detected.

Cytoplasm – the protoplasm of the cell which is found outside the nucleus.

Decomposition – the process of rotting or breaking down.

Diffusion – movement of particles from high to low concentration down a concentration gradient.

Dimer – a chemical structure formed from two subunits.

Dislocation – the displacement of a part, especially the displacement of a bone at the joint.

DNA (deoxyribonucleic acid) – the nucleic acid molecules that make up chromosomes and carry genetic information; found in every cell of every organism; control cell chemistry.

Ecosystem – the living biological and non-living physical components of the environment.

Effector – the part of the body, e.g. a muscle or a gland, which produces a response to a stimulus.

Embryo – a ball of cells that will develop into a human / animal baby.

Enhanced global warming – the effect of increasing levels of carbon dioxide in the atmosphere.

Enzyme – a protein molecule and biological catalyst found in living organisms that helps chemical reactions to take place (usually by increasing the rate of reaction).

Evolve – to change naturally over a period of time.

Extinct – a species that has died out.

False negative – when a genetic test incorrectly states that the situation is normal (suggesting that the disease-causing allele is absent) when it is in fact positive.

False positive – when a genetic test incorrectly states that the result is positive (and therefore the disease-causing alleles are present) when in fact it is negative.

Fertilisation – the fusion of a male gamete with a female gamete.

Fetus – an unborn animal / human baby.

Food chain – a representation of the feeding relationship between organisms; energy is transferred up the chain.

Food web – the graphical representation of all the linked food chains in an ecosystem, allowing the identification of feeding relationships.

Fossil fuel – fuel formed in the ground, over millions of years, from the remains of dead plants and animals.

Fuel – a substance that releases energy when burned in the presence of oxygen.

Gamete – a specialised sex cell.

Gene – a small section of DNA, in a chromosome, that determines a particular characteristic; controls cellular activity by providing instructions (coding) for the production of a specific protein.

Genetic test – a test to determine if an individual has a genetic disorder.

Genetic modification – the change in the genetic make-up of an organism.

Genetic screening – the process of testing to see if certain genes are present.

Genomics – the study of the genomes of organisms.

Heart – a muscular organ that pumps blood around the body.

Heterotroph – an organism that is unable to make its own food; consumes other organisms.

Homeostasis – the maintenance of constant internal conditions in the body.

Hormone – a chemical messenger, made in ductless glands, which travels around the body in the blood to affect target organs elsewhere in the body.

Immunity – the individual is protected against infection from a specific microorganism by their immune system.

Intensive farming – a method of farming that uses artificial pesticides and fertilisers and controlled environments to maximize food production.

IVF (*In vitro* fertilisation) – a technique in which egg cells are fertilised outside the woman's body.

Ligament – the tissue that connects a bone to a joint.

Meiosis – the type of cell division that forms daughter cells with half the number of chromosomes as the parent cell; produces gametes.

Meristem – an area where unspecialised cells divide, producing plant growth, e.g. roots and shoots.

Mitochondria – membrane-bound organelles containing enzymes that carry out respiration for a cell.

Mitosis – the type of cell division that forms two daughter cells, each with the same number of chromosomes as the parent cell.

Muscle – tissue that can contract and relax to produce movement.

Nanotechnology – the application of technology on scales of 1 to 100nm in at least one dimension.

Glossary

Natural selection – the process by which organisms that are better adapted to their environment are able to survive and reproduce, passing on their characteristics to their offspring.

Neuron – a specialised cell that transmits electrical messages (nerve impulses) when stimulated.

Nitrification – the conversion of ammonia to nitrite and then to nitrate by microorganisms.

Non-biodegradable – a substance that does not decompose naturally by the action of microorganisms.

Non-renewable resources – resources (especially energy sources) that cannot be replaced in a lifetime.

Nucleus – the membrane-bound organelle where DNA is stored. It is the control centre of a cell.

Organelle – a specialised subunit of a cell with its own function.

Osmosis – the net movement of water from a dilute solution (lots of water) to a more concentrated solution (little water) across a partially permeable membrane.

Pesticide – a chemical used to destroy insects or other pests.

pH – a measure of acidity or alkalinity.

Photosynthesis – the chemical process that takes place in green plants where water combines with carbon dioxide to produce glucose using light energy.

Phototropism – a plant's response to light.

Phylogenetic tree – a graphical representation of the evolutionary relationships between living things based on analysis of DNA.

Physiotherapist – a specialist in the treatment of skeletal-muscular injuries.

Placebo – an inert (non-effective) dummy pill or treatment.

Plasma – the clear fluid part of blood that contains proteins and minerals.

Plasmid – a section of DNA in a bacterium that reproduces independently of the main chromosome.

Platelet – a tiny particle found in blood plasma.

Receptor – the part of the nervous system that detects a stimulus; a sense organ, e.g. eyes, ears, nose, etc.

Recycling – to reuse materials that would otherwise be considered waste.

Reflex action – an involuntary action, e.g. automatically jerking your hand away from something hot; a fast, involuntary response to a stimulus.

Respiration – the liberation of energy from food.

Ribosome – a small structure found in the cytoplasm of living cells, where protein synthesis takes place.

Selective breeding – the process by which animals are selected and mated to produce offspring with desirable characteristics (artificial selection).

Sexual reproduction – reproduction involving the fusing together of gametes formed via meiosis.

Specialised – developed or adapted for a specific function.

Species – a group of organisms capable of breeding to produce fertile offspring.

Sprain – a stretch or tear in a ligament.

Stem cell – a cell that can give rise to specialised cells.

Stimulus – a change in the environment that can elicit a response in a living organism.

Sustainability – the ability to retain a variety of living organisms and resources over time.

Symptom – a visible or noticeable effect of a disease, illness or injury.

Synapse – the small gap between adjacent neurons.

Tendons – tissue that connects a muscle to a bone.

Glossary

Vaccine – a liquid preparation used to make the body produce antibodies to provide protection against disease.

Variation – the differences between individuals of the same species.

Vector – an organism (often a microorganism) used to transfer a gene, or genes, from one organism to another.

Vein – a type of blood vessel that transports blood towards the heart.

X-ray – an imaging technique that produces shadow pictures of bone and metal.

Zygote – the first cell formed after the fertilisation of an egg by a sperm.

HT

Active transport – the movement of substances against a concentration gradient; requires energy.

Auxin – a plant hormone that affects the growth and development of the plant.

Bioaccumulation – the accumulation of chemical substances in a food chain such that the concentration increases with each trophic level.

Denatured – the state of an enzyme that has been destroyed by heat or pH and can no longer work.

Denitrification – the breakdown of nitrates into atmospheric nitrogen by microorganisms.

Detritivore – a heterotroph that helps break down dead or decaying material (detritus).

Epidemic – a disease that occurs suddenly in numbers that are much higher than normally expected.

Eutrophication – a situation in which the nutrient content is enriched to the point where the productivity of an ecosystem is excessively increased.

Genotype – the genetic make-up of an organisms (the combination of alleles).

Homozygous – both alleles present in a pair are identical.

Heterozygous – each allele in a pair is different.

Hypothalamus – a part of the brain responsible for maintaining homeostasis.

Mutation – a spontaneous change in the genetic material of a cell.

Negative feedback – a process in which a signal from a receptor instructs an effector to reverse an action.

Nitrogen fixation – the breakdown of nitrogen gas into nitrate by microorganisms.

Phenotype – the observable characteristics that the organism has.

Pituitary gland – the small gland at the base of the brain that produces hormones.

Pre-implantation diagnosis – the method of testing a cell from an eight cell embryo to determine if it carries genetic disorders.

Primer – a strand of nucleic acid that serves as a starting point for DNA synthesis.

Vasoconstriction – the narrowing of blood vessels to reduce blood flow.

Vasodilation – the widening of blood vessels to let more blood flow.

Index

Active sites 33–34
Active transport 36
Adaptations 20
ADH 19
Aerobic respiration 38
Alcohol 19, 40, 62
Alleles 4, 5
Anaerobic respiration 38, 40
Androgen 6
Animal cells 39
Antibiotics 13, 14, 79
Antibodies 12–13
Antigens 12
Antimicrobials 13–14
Arteries 16, 67
Asexual reproduction 10
Auxins 43
Axons 52

Bacteria 11, 14, 39, 79
Bioaccumulation 76
Biodiversity 28
Biomedical engineering 83
Blood 12, 67–68
Blood pressure 17, 19
Blood sugar 70–71
Body fat 63
Body mass index 63
Body temperature 69–70
Brain 52, 54, 56, 69

Cacti 20
Capillaries 16, 68
Carbon cycle 23
Cartilage 61
Cells 32, 39, 41
    animal 39
    division of 41, 45–46
    microbial 39
    plant 35
Central nervous system 52
Cerebral cortex 54, 56
Child development 56
Chlorophyll 34
Chromosomes 3–4, 6
Circulatory system 66
Classification 28
Clinical trials 14–15
Clones 10, 48
Closed loop systems 72, 73
Combustion 23
Competition 20, 74
Crude oil 29, 78
Cystic fibrosis 6, 7

Decay organisms 22
Decomposers 22

Deforestation 77
Dehydration 19, 69
Denaturing 33
Desertification 77
Detritivores 22
Diabetes 71
Diffusion 35
Digestive enzymes 72–73
DNA 3, 25, 45, 46–47, 81–82
Drugs
    effect 19, 53
    misuse 17, 53
    testing 14–15

Ecosystem services 75
Ecosystems 72–75
    human activity and 75–78
Effectors 50, 51
Energy
    decay organisms and 22
    efficiency 22
    in food chains 21, 32
    in living things 32
Enzymes 32–34, 72–73, 79
Eutrophication 77
Evolution 25, 28
Exercise
    developing regime 61–62
    effect 62
    importance of 71
    injuries 64–65
Extinction 29
Eyes 3, 51

Fermentation 40, 79
Fertilisation 4, 41–42
Fieldwork 37
Food chains 21
Food webs 21
Fossil record 25
Fungi 11

Genes 3–5, 25, 26, 48
Genetic diagrams 5
Genetic disorders 6–9, 81
Genetic modification 80
Genetic testing 8–9, 81–82
Genotype 3, 5
Glucose
    in the body 70
    in plants 34–35
Glycemic index 71
Growth 45

Heart 16, 66–67, 83
Heart disease 16–17
Heat stroke 69

Homeostasis 18–19, 69–70
Hormones
    in the body 50, 51, 70
    in plants 43
Huntington's disease 6, 7
Hypothalamus 69

Immune system 12–13
In vitro fertilisation 8
Indicators 24
Industrial fermentation 79
Infectious diseases 11
Insulin 70, 71, 79, 80

Joints 60–61

Kidneys 18–19

Ligaments 61, 65
Lock and key model 32

Magnetic resonance imaging 54
Meiosis 46
Memory 57
Meristems 42, 43
Messenger RNA 47
Microorganisms 11–12
Mitosis 41, 45
Muscles 51
Mutations 26

Nanotechnology 82–83
Natural selection 26, 27, 28
Negative feedback 18
Nervous system 49–53
Neurons 50, 52, 56
Nitrogen cycle 23

Osmosis 36

Packaging 29
Peripheral nervous system 52
PH, enzymes and 33, 34
Phenotype 3, 5
Phloem 43
Photosynthesis 23, 32, 34, 36
Phototropism 43–44
Placebos 15
Plants
    cell structure 35
    glucose in 34–35
    meristems 42, 43
    movement of substances 35–36, 43
    photosynthesis 23, 32, 34, 36

phototropism 43–44
reproduction 10, 74
soil and 75
Plasma 68
Plasmids 80
Platelets 68
Pre-implantation genetic diagnosis 8
Punnett square diagrams 5

Quadrats 37

Receptors 49, 50, 51
Red blood cells 68
Reflexes 49, 50–51, 55
Reproduction 10, 41–42, 46, 74
Respiration 16, 23, 32, 38, 40

Selective breeding 26–27
Serotonin 53
Sex cells 4
Sexual reproduction 10, 41–42, 46
Side effects 13
Soil 75
Species 20
Sphygmomanometer 17
Spinal cord 50, 52, 83
Spinal reflex arc 50–51
Sprains 64–65
Stem cells 10, 42, 48, 83
Sustainability 29
Synapses 53, 56
Synovial fluid 61

Temperature
    body 69–70
    enzymes and 33
    photosynthesis and 36
Tendons 61, 65
Tissue fluid 68
Tissues 41

Vaccination 13
Variation 3, 4
Vasoconstriction 70
Vasodilation 70
Veins 16, 67
Viruses 11, 80

Water balance 18–19
White blood cells 12, 68

Xylem 43

Yeast 39, 40